LEARNING TO SEE DATA

How to Interpret the Visual Language of Charts

BEN JONES
Founder and CEO, Data Literacy

DATA LITERACY PRESS

ISBN: 978-1-7332634-4-3 (paperback)
ISBN: 978-1-7332634-5-0 (ebook)

Printed in the United States of America.

For my mother, Jane,
who first taught me to see clearly.

MORE BY BEN JONES

The Data Literacy Series

Data Literacy Fundamentals

Learning to See Data

Read, Write, Think Data

Other Books

The Introspective Entrepreneur

Avoiding Data Pitfalls

Communicating Data with Tableau

CONTENTS

This book is a companion to the Data Literacy Level 1 online course which can be found at **https://dataliteracy.com/data-literacy-level-1/**.

PREFACE

Data visualization is a passion of mine. I've seen it open people's eyes and reveal to them truths they never imagined. I've seen it make people laugh at a witty analogy, gasp at an outrageous outlier, or scratch their heads at an unexpected trend. And I've seen the ability to read and interpret charts go from a rare skill of the specialist to a requirement on just about every job description you can find.

But I've also seen data visualizations cause people to scratch their heads in confusion. I've seen people turn away from charts feeling intimidated and even embarrassed. And I've seen people jump through all manner of hoops in an attempt to discredit them. Charts can empower us, but they can also bewilder us and inconvenience us.

I didn't really intend to have a career in data visualization, and my plan wasn't to become an instructor or trainer. I started out as a young mechanical engineer, fresh out of UCLA, and ready to design nifty electromechanical gadgets. I did that for a while and loved it. I redesigned an elliptical machine to have a wider stride pattern. I took a sensor technology from missiles and used it to enable a diesel truck to vary its level of turbocharge. I troubleshot defective insulin pumps and discovered the cause of a mysterious battery failure: tin-plated connectors that were prone to rust.

All of these engineering projects involved data, and more often than not they also involved charts and graphs that let me see what was going on in the data. As great as these jobs were, they weren't really paying me enough to support a budding family in Southern California where we lived. So I got swept up into the "Six Sigma" movement in 2005, complete with a title inspired by the martial arts. At that time Jack Welch was still widely admired and even revered, as some of the costlier side effects of his data-first managing style were not yet common knowledge.

Having traded in my junior engineer's badge for a fancy "black belt" that my father never did understand, I started working with process owners to eliminate waste and reduce defects that were costing the company a lot of money. As an internal consultant with at *least* two hours of "soft skills" training, I became a hero or a scourge, depending on who you talked to.

For all of the hokeyness of the training waves and the various colored belts and the gold star project trophies, we actually did save the company a lot of money. I believe that. And I became aware of the fact that, when visualized properly, data has the power to change minds and to change the future.

I became hooked on the topic. I subscribed to every data blogger's RSS feed I could find. People like Andy Kirk and Nathan Yau were idols to me, and I dreamed of finding a way to make my passion a full-time endeavor. It seemed like such a long way off. So I did what any hopeless dreamer would do: I started my own blog. I called it DataRemixed. Hello world!

A mere 12 months later, I was getting recruited by the likes of Facebook and ESPN, and I couldn't believe it. I was about to make a big leap. But, since my career dreams were somehow coming to life, I figured I'd swing for the fences and see if Tableau was interested in hiring me. They were and they did. Instead of Silicon Valley or Bristol, Connecticut, Seattle became my new home. Tableau Public was my jam, and I felt like I was doing the exact thing I was supposed to be doing. I hope everyone gets to feel that way sometime in their career.

Almost ten years, five books, and thousands of students later, I can say that teaching data and writing about it is my thing. I've branched out to start Data Literacy, along with my wife, Becky, and I really love it. I feel very blessed. And we're only getting started. This is the end of our second year running Data Literacy, and we're beginning to make some serious headway. Making it through a pandemic at the helm of a brand-new business isn't for the faint of heart. I couldn't have done it without my partner.

This book is for my mom, Jane. She's "a real hoot," as she'd say, and I guarantee you'll like her if you ever meet her. The title, of course, is *Learning to See Data*, and when I thought about who I wanted to dedicate this book to, I asked myself who taught me to see things. If I think back far enough, she was the first one to do that. She didn't teach me to see data as much as she taught me to see the best in others and in myself. I don't think I'd be as good at what I do if that hadn't been the case.

So thanks, Mom. "Oo-de-lally," I sure do appreciate you.

Ben Jones
Bellevue, Washington
November 20, 2020

INTRODUCTION

"There is quite a difference between simply looking at a chart and seeing it."
—Mary Eleanor Spear

When you come across a chart, a map, or a data dashboard as you go through your day, what's the first thing you think or feel? Does the display pique your curiosity, making you want to know what the visual can teach you about the world around you? Or does the display evoke a feeling of trepidation or even panic due to a concern that it might not make any sense to you?

Many of us still have the second reaction, and that's a problem. As we all can attest, we're inundated with displays of data in the three domains of application—professional, public, and personal—that we covered in the previous book and course, *Data Literacy Fundamentals*.

At work, we see charts and graphs about our industry or about the performance of our company or our team. In society, we come across dashboards conveying critical information about public health matters or election results. And in our own personal lives, we look at visuals related to our finances or our health and fitness on a regular basis.

But how do we use our visual cortex to wrap our minds around what these charts are actually telling us? And how are they built in the first place? What's the difference between a bar chart and a pie chart and what do the lines and legends and shapes and colors *mean*?

The overarching goal of this book, and the course that goes along with it, is to leave you feeling like you have what it takes to read and interpret the visual displays of data that come into your view as you go about your business and your life.

In order for us to achieve this together, we'll need to tackle the topic from two different angles: from the theoretical angle and from the practical angle. The theoretical angle builds your knowledge of concepts and principles of graphical representation of data. The practical angle hones your level of skill and comfort interpreting various types of charts and graphs, noticing and extracting the insights that they contain. Both angles are critical.

This book is divided into eight chapters that each focus on a different type of *perspective* that we can learn to develop about our data, such as part-to-whole or change over time. Each chapter features its own set of common chart types that convey these different perspectives. The charts included in the book aren't by any means exhaustive of all possible chart types, but the most critical ones are included, and even some frivolous ones, too.

You'll also come across this symbol from time to time in the charts and diagrams included in the book:

The "Data Pitfall Alert" Sign

This "data pitfall alert" sign provides a warning about misleading aspects of charts that commonly lead their readers astray. Part of being a good traveler is knowing where *not* to go, and this sign will be your clue to those dips and swerves in the road that you'll want to navigate carefully. If you're interested in learning more about the common blunders that we all make when working with data, get yourself a copy of *Avoiding Data Pitfalls*.[1]

A quick word about some sources of inspiration that we drew from as we created this course. The visual language of data is more important than ever, but we know that it's certainly not new. Pioneers of old like William Playfair, Florence Nightingale, W.E.B DuBois, and many others created this visual language and developed it over time. In the past few decades, researchers have studied it and software developers have built very powerful tools to help us put it to use at a whole new level.

Among the many talented people in the field of data visualization that have contributed to our understanding over the years, three women in particular stand out as sources of knowledge and inspiration for this book.

First, Mary Eleanor Spear was an American visual information specialist who worked for various U.S. federal agencies for many decades, starting in the 1920s. She wrote two books that haven't really been given their full due: *Charting Statistics*, which she wrote in 1952, and a reboot, *Practical Charting Techniques*, which was published in 1969.

I had a chance to speak with her granddaughter recently, and she told me about her grandmother's enthusiasm for her craft, and the attention to detail she paid to the handmade charts she painstakingly created in an upstairs office next to her bedroom in her home in Takoma Park, Maryland. The inspirational quote for this course comes from page 63 of Spear's book *Practical Charting Techniques*:[2]

> *Learn to see details. There is quite a difference between simply looking at a chart and seeing it. Looking is your first visual impression, while seeing involves the studying of distinct parts of the visual.*

A second source of inspiration, Barbara Tversky, is an emerita professor of psychology at Stanford University and a professor of psychology

1. Jones, B. 2018, *Avoiding Data Pitfalls*, John Wiley & Sons, Inc. New Jersey
2. Spear, M E 1969, *Practical Charting Techniques,* McGraw Hill, Inc.

at Teachers College at Columbia University. Her recent book, *Mind in Motion*,[3] explains in clear terms the experiments that she and her fellow researchers carried out over the years. Many of the eye-opening insights inform our understanding of how our brains make use of objects in our field of view, including the graphical representations we cover in the course.

Third, and most importantly, Tamara Munzner is an American-Canadian scientist and an expert in information visualization who works as a professor of computer science at the University of British Columbia. Munzner's textbook *Visualization Analysis and Design*[4] is a tour de force in the field of visualization, and a must-read for anyone looking to develop a deep understanding of this critical field. We make use of wonderful graphics that she and the illustrator of her text, Eamonn Maguire, thoughtfully designed and generously shared with the world via Creative Commons. There's no other source that explains in as clear and thorough a way the critical underlying concepts of graphical marks and visual encoding channels.

Okay, let's get started! Broad fluency in the visual language of data still isn't where it needs to be. Most adults didn't get a formal education that included a focus on chart interpretation, and we know that many do struggle to grasp the meaning and message of basic charts like line charts or dot plots, let alone more complex charts like treemaps or box-and-whisker plots.

With this book we aim to address this urgent issue, and to help close the education gap of our era, one reader at a time, starting with you. Good luck, and enjoy the journey!

3. Tversky, B 2019, *Mind in Motion*, Basic Books, New York
4. Munzner, T 2014, *Visualization Analysis and Design*, A K Peters/CRC Press, Boca Raton

Chapter 1

ENCODING DATA IN GRAPHICAL FORM

"The visual system provides a very high bandwidth channel to our brains."
—Tamara Munzner

Most of us come across graphical representations of data on a weekly or even daily basis. Often what we see is relatively straightforward and easy to understand, such as a bar chart showing our company's sales by quarter. Other times what we see is more complex than that, such as a multifaceted dashboard of sales by product and industry, and we find ourselves struggling to make sense of it. Sometimes even seemingly simple displays can be confusing.

In this book and the course it accompanies, we will focus on the role of the *interpreter* of data visuals as opposed to the *creator* of such visuals. Both creator and interpreter are responsible for their part of the information exchange that revolves around these ubiquitous data visuals. We'll consider the role of the creator more closely in the subsequent book and course that deals with the process of converting raw data into insight and wisdom.

For now, though, let's put ourselves in the shoes of the receiver of the signal—the chart reader or interpreter. The more time goes on, the more frequently we encounter data visuals, both simple and complex, and the more important it is that each of us know how to make sense of them.

The Human Visual System

In order for us to become fluent readers and interpreters of visual displays of data, we need to have our minds opened to the way the mind itself works when it encounters such displays. Our visual system is a magnificent element of the central nervous system, and it allows us to quickly recognize patterns and to process information about objects within our field of view. As the epigraph of this chapter relates, we receive an incredible amount of information that way.

We briefly considered the visual system in the previous book and course of this series, *Data Literacy Fundamentals*.[5] But the human visual capability becomes our central focal point (pun intended) of this second book and course on reading and interpreting visual data.

To revisit, light waves bounce off of objects or emanate from sources around us and pass into the eye through the cornea and lens, which refract the light and shine it onto the back of the retina, forming an image there. The rods and cones of the retina then turn this image into electrical impulses, which are carried by the optical nerve through the lateral geniculate nucleus (LGN) in the thalamus—our sensory information relay station—and into the visual cortex of the brain, which is located all the way in the back of the skull in the occipital lobe.

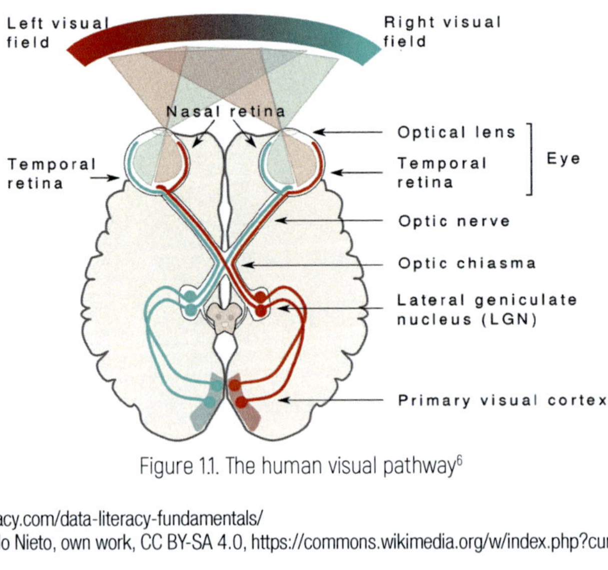

Figure 1.1. The human visual pathway[6]

5. https://dataliteracy.com/data-literacy-fundamentals/
6. By Miquel Perello Nieto, own work, CC BY-SA 4.0, https://commons.wikimedia.org/w/index.php?curid=37868501

Figure 1.1 shows the various parts of the human visual system, as well as the direction of flow of the light from both the left and right sides of the field of view and the resulting electrical signals that pass into the brain.

This process results in sighted individuals becoming consciously aware of the shapes, colors, textures, groupings, and movements of objects within their environment. Our conscious mind can also use the visual information we receive to answer questions about what we see, such as:

- How many distinct objects do I see?
- What similarities and differences can I notice about these objects?
- Are there any groups or clusters of objects?
- Can I find an object with a particular characteristic, like size, shape or color?
- Which objects are {longer, larger, more tilted, darker, moving faster, etc.} than other objects?
- How much {longer, larger, more tilted, darker, faster, etc.} is one object compared to another?

As amazing as this capability is, our visual cortex isn't a completely flawless engine of interpretation. It can fall prey to optical illusions, such as in Figure 1.2, in which the two lines running through the center of the image appear to be bowed or curved in the center, when in fact they are actually horizontal and perfectly parallel. Feel free to use a ruler or straightedge to verify for yourself.

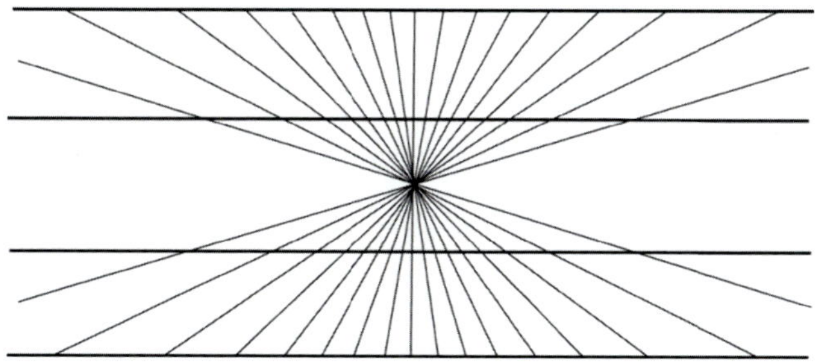

Figure 1.2. An optical illusion of straight lines that seem to curve[7]

7. http://www.publicdomainfiles.com/show_file.php?id=13921991421195

Additionally, as we'll see in this chapter, it turns out that we are able to carry out some visual tasks and comparisons more easily and effectively than others. For example, we tend to judge the difference in length of two straight lines more accurately than the difference in area of two circles or the volume of two spheres.

This is highly relevant to our present study, because these quirks and variances in our visual capacities apply to the way we read and interpret charts, maps, and dashboards that we encounter every day. By understanding which tasks tend to be easier and which tend to be more difficult and error prone for us, we can avoid two common pitfalls: first, the pitfall of misinterpretation, and second, the pitfall of self-blame. People often misinterpret charts, and they often feel that there's something wrong with them when they do. But there's a good chance that the difficulties they experienced are ones that almost all of us would share based on how the human brain works.

The fact is that there's no guarantee that the chart we're looking at will enable us to get the information we need from it, depending on what it includes and how it has been built. It might have been built poorly, or it might have been built to help someone carry out a different mental task. In this respect, the details of the chart can matter a great deal.

Every one of us will struggle to answer our question when given the "wrong" chart. This doesn't necessarily mean that the given chart is completely useless; it might just be poorly suited for our particular question or task. So what do we do in those situations?

Well, in order to know how to move forward, we need to be able to look closely at the different elements of visuals we see and understand how these elements have been put together. Once we've done that, we can assess whether or not the resulting visual can give us what we need or not.

It's important to note that we shouldn't expect every chart to answer all of our questions about a particular topic. That's just not practical. In the rest of this book and course, we'll bring to life a variety of different chart types, and we'll explore their relative strengths and weaknesses as tools or visual aids for making sense out of data. Understanding these concepts of visual data and acquiring these skills of interpretation allows us to put in place a key cornerstone of the foundation of data literacy: reading charts well.

A Review of Data Scale Types

Before we compare the ways in which data can be encoded in visual form, let's start by reviewing the different types of measurement scales into which the data variables can fall. When we can think clearly about the data itself, we can begin to think clearly about how it can be visualized.

In the fourth section of *Data Literacy Fundamentals*, we introduced four different data scale types: nominal, ordinal, interval, and ratio. These four data scale types were proposed in 1946 by Stanley Smith Stevens, a Harvard psychologist and researcher who was studying the way we as humans perceive different sensory inputs such as sound.[8] Since the four scale types formed the acronym N-O-I-R, we chose to remember them as a bottle of pinot noir, a type of red wine:

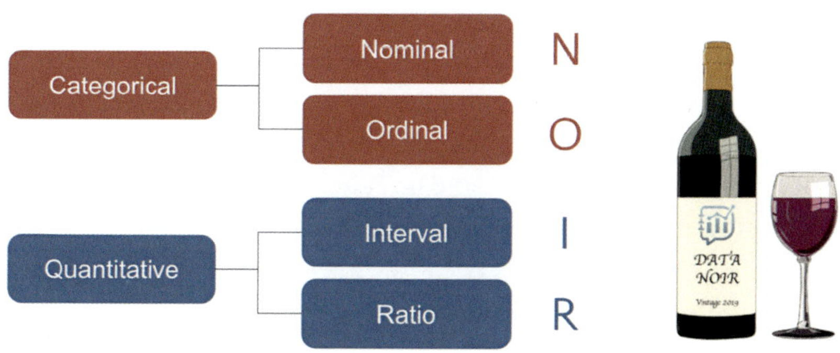

Figure 1.3. The four types of data scales proposed by S.S. Stevens

To review, the first scale in the typology, the nominal scale, is one of the two types of categorical variables. With nominal scales, the levels of the scale don't have an inherent order or rank to them. Some examples are names, gender, country of birth, and hair color, to name just a few. Nominal scale variables can simply tell us whether two items are the same or different in some respect.

Ordinal scales are also categorical, and therefore they also tell us whether two items are similar in some way. But unlike nominal scales,

8. Stevens, S. S. "On the Theory of Scales of Measurement." *Science*, vol. 103 (no. 2684, 1946): 677–680. *JSTOR*, www.jstor.org/stable/1671815. Accessed 12 September 2020.

the various levels of ordinal variables do have an inherent order to them:

- Which medal did an athlete win in a tournament (gold, silver, bronze)?
- Which T-shirt size should I order (small, medium, large, x-large)?
- What rating would I give to a product I just bought (1-5 stars)?

By their very definitions, the levels of these scales belong in a specific order, with one end ranking higher and the other end ranking lower than the other. For this reason, we can also use ordinal scales to compare whether a particular entity is greater than, equal to, or less than another entity. A gold medal is greater than a silver medal, which is greater than a bronze one.

Interval and ratio scales are considered quantitative because they are measured and conveyed by numbers in which the intervals between numbers are consistent. What does this mean? Well, consider the difference between product rating (1–5 stars) and shipment quantity in units (say, 1–5 boxes). With product rating, which is an ordinal scale variable, the difference between a 1-star and a 2-star rating isn't necessarily the same as the difference between a 4-star and a 5-star rating, even though both gaps are 1-star. Why not? Because the way we assign ratings based on our experience and preference is subjective and not perfectly linear. One person might be generous and give 5-star ratings to most products that are at least decent, while another person might only give a 5-star rating only to products that really blow them away.

On the other hand, the numerical values that shipment quantity can take are linear and consistent and therefore the scale is quantitative. How are the values of shipment quantity linear and consistent? The difference between a shipment with one box and a shipment with two boxes is the exact same one box difference as between a shipment with four boxes and a shipment with five boxes:

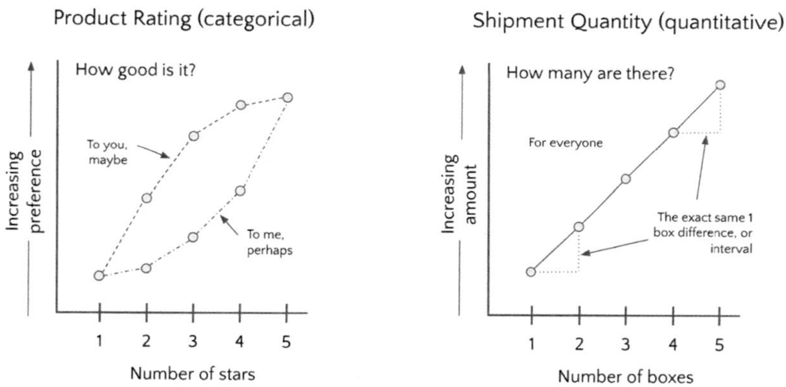

Figure 1.4. Quantitative scales preserve equality of intervals

This important difference between categorical and quantitative scales will become critical as we consider later in this chapter how these variables can map to visual traits such as length, size, and color.

As shown in Figure 1.3, there are two types of quantitative scales: interval and ratio. The difference between interval and ratio scales lies in the all-important meaning of the value zero. Interval scales have a zero that is arbitrary or a matter of convention, such as 0°F or 0°C. Zero on both the Fahrenheit and the Celsius temperature scale doesn't correspond to "no temperature," or total lack of movement of molecules in the air. In other words, these aren't "true" or "absolute" zero points.

In contrast, returning to our shipment quantity variable, if we ship 0 boxes of a product, then that means the shipment doesn't have *any* boxes of that product. None. Because of this fact, shipment quantity has a "true" zero and is a ratio scale.

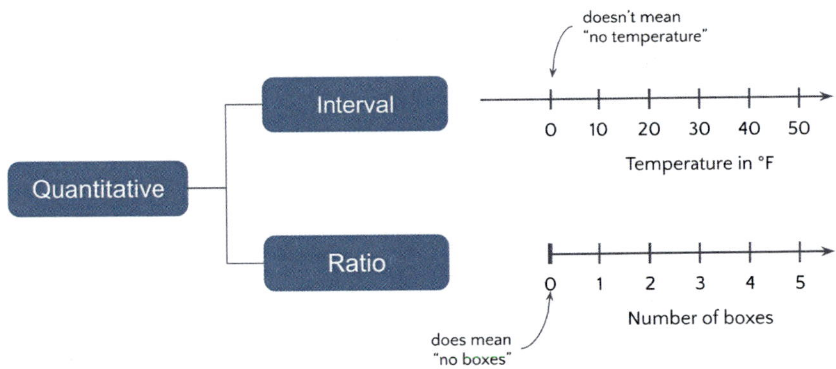

Figure 1.5. Presence of a "true" or "absolute" zero is the difference between ratio and interval

Why does this difference matter, though, as we think about graphical displays of data? As we'll see, it matters a lot when we consider at what value the axes should start, or how to interpret relative proportions between objects in the view. More on this point soon.

In *Data Literacy Fundamentals*, we also stressed that Stevens's four-scale typology isn't exhaustive. For example, his typology doesn't address important questions, like whether or not a quantity can take a negative value, or whether the quantity is discrete (only able to take a finite number of values) or continuous (theoretically able to take an infinite number of values between two limits).

Additionally, Stevens's typology shouldn't be applied too rigidly when we explore or analyze data. Often we need to be flexible in how we treat the variables in our data, and the most suitable scale type for a particular variable can sometimes depend on the questions we ask of it and the usefulness of the resulting analysis. Why tie our hands if doing so doesn't serve us well?

This way of thinking—namely that we can come up with general principles, guidelines, and rules of thumb as opposed to rigid laws and immutable edicts—applies to many aspects of working with data, as we'll see throughout this book and course.

Graphical Marks and Encoding Channels

Now that we have reminded ourselves about the different scale types that our data variables can fall into, let's consider how people use these variables to create visuals that can show us what the data contains. Instead of just jumping right into consideration of common chart types, let's first consider the basic building blocks of all data visuals: marks and channels.

In short, **marks** are the graphical elements in the chart, and **channels** are attributes of these marks that can be controlled by the data. Jeffrey Heer, computer science professor at the University of Washington, summarizes these two terms concisely:

> *A visualization represents data using a collection of graphical marks such as bars, lines, and point symbols. The attributes of a mark—such as its position, shape, size, or color—serve as channels in which we can encode underlying data values. With a basic framework of data types, marks, and encoding channels, we can concisely create a wide variety of visualizations.*[9]

As Tamara Munzner of the University of British Columbia relates in her groundbreaking textbook *Visualization Analysis and Design,*[10] some examples of marks that we can find within visual displays of data include points (0D), lines (1D), areas (2D), and even the less-common volumes (3D):

Figure 1.6. Munzner's diagram showing different types of marks

9. https://observablehq.com/@uwdata/data-types-graphical-marks-and-visual-encoding-channels
10. Tamara Munzner, *Visualization Analysis and Design*, with illustrations by Eamonn Maguire (A K Peters Visualization Series, CRC Press, 2014).

As for visual channels that can modify how these marks appear to us, Munzner provides the following helpful diagram that outlines a few of the common visual channels: position (horizontal, vertical, and both), shape, color, tilt or angle, 1D size (length), 2D size (area), and 3D size (volume):

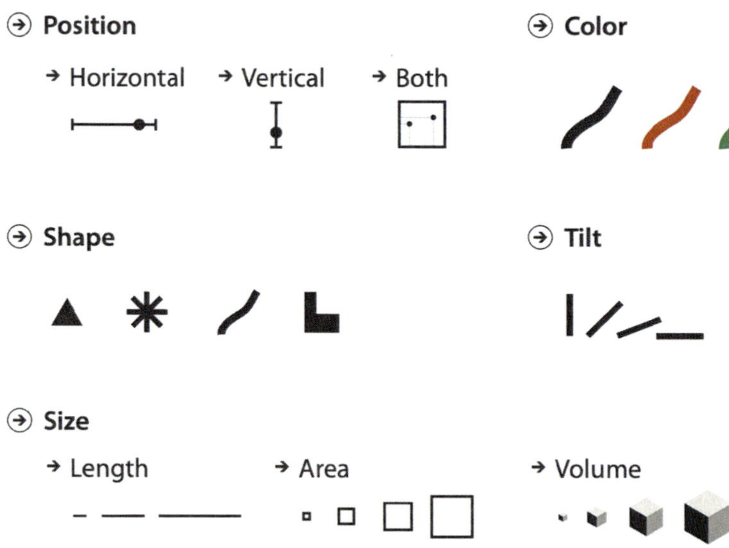

Figure 1.7. Munzner's diagram showing different types of visual channels

In order to really understand how marks and channels combine to create the charts and graphs we see all around us, let's consider an example using real-world data. We'll consider a simple data set contained in a small table that we encountered in the previous book and course, the population and gross domestic product (GDP) of the three most populous countries of the world in 2018: China, India, and the United States.

Country	Income Group	Year	Population (in billions)	GDP (in trillions of USD)
China	Upper middle	2018	1.393	$13.61
India	Lower middle	2018	1.353	$2.72
United States	High	2018	0.327	$20.54

Figure 1.8. A simple country table with five variables

The five variables in this table, each in their own column, account for all four of the scale types we just reviewed. Country is a nominal variable since there's no inherent order to the countries. Income Group is ordinal since the levels of this categorical variable belong in a specific order of descending wealth: high, upper middle, lower middle, and finally low. This fourth and last level of Income Group doesn't actually appear in this small table since there aren't any countries included from this lowest of income groups. Year is an interval scale quantity, since there is no such thing as an absolute zero year that corresponds to "no year" or "absence of time." Finally, Population and GDP are ratio scale variables since they're quantities with a true zero—a population of zero would correspond to no people living in the country, and an economy with a GDP of $0 wouldn't have any output at all.

But how can we use marks and channels to encode the data in this table? Let's see how the table of numbers can be used to create a simple bar chart, step-by-step.

a. First, in Figure 1.9a, the marks are bars, and the lengths of the bars are encoded using the GDP variable. The x-axis at the bottom makes this encoding clear to us.

b. Then, in Figure 1.9b, we change the mark type from bars to points, or filled circles, and we encode the (horizontal) x-position of the circles using GDP.

c. Next, in Figure 1.9c, we use the color channel of the circles to communicate the Income Group variable: the higher the income group, the darker the green.

d. Finally, in Figure 1.9d, we use the area channel (2D size) of the colored circles to encode the population of each of the three countries in 2018.

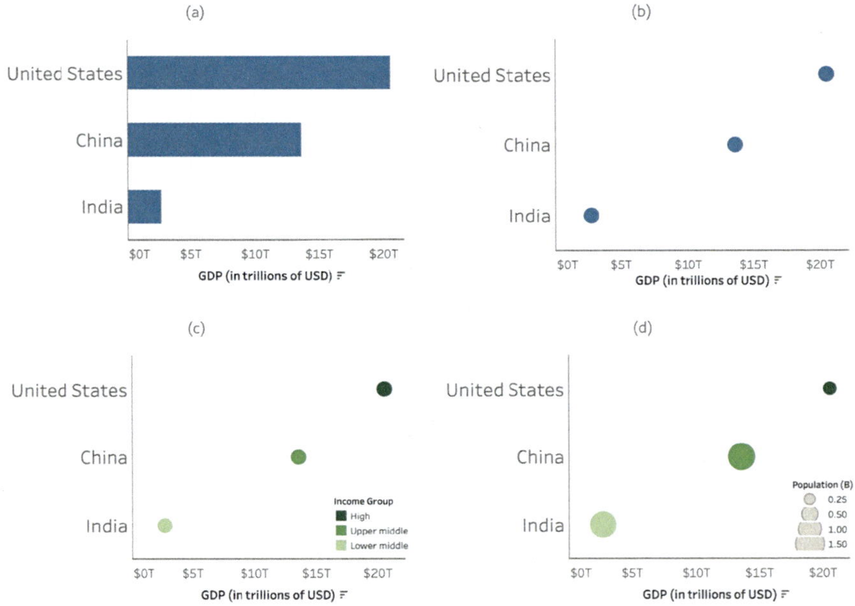

Figure 1.9. Beginning to use marks and channels to visualize tabular data

In this simple example, we can see that different marks can be used to represent our data, and both categorical and quantitative variables can be used to control, or encode, different visual channels to help the reader of the charts be able to glean interesting information and knowledge from what they see.

But so what? Why was this helpful? It illustrates why graphical representations of data can be powerful tools for us to use when we seek to make sense of data. In a nutshell, charts can serve as a shortcut for the human mind to convert data into knowledge. When we look at the final version of the visual in Figure 1.9d, we notice right away that even though the population of the United States is much smaller than either China's or India's population, its gross domestic product was quite a bit higher than either of these larger countries in 2018. And we can also easily see that even though China and India had very similar population sizes in 2018, China's GDP was much greater than India's that year.

And while we could've gleaned this knowledge just by looking at the table, it's likely that it would have taken longer to get there. The insight wouldn't have jumped out at us like it did when we looked at the chart. And what are we to do if the table includes dozens, or hundreds, or even millions of rows? In those cases, visual displays of data are wonderful shortcuts to knowledge.

If we revisit the DIKW Pyramid covered in the *Data Literacy Fundamentals* book and course, we see that the charts in this example provide a pathway up the levels of the pyramid so that we can more quickly convert data into knowledge. At the first level of the pyramid, the raw data was collected and made available for us by the World Bank.[11] To get to the second level, we turned the raw data into information by interpreting the raw values in the table using the units of measure provided in the column headers. Then we moved up to the third level by making fast associations using the marks and channels of the visual, converting information into knowledge.

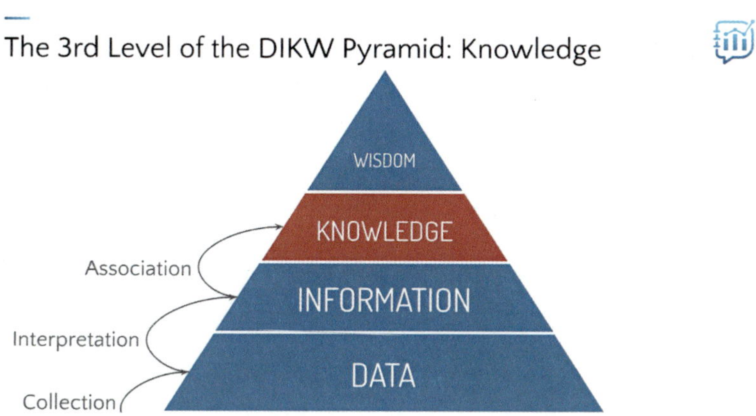

Figure 1.10. Getting to the Knowledge level of the DIKW Pyramid

In this first real-world example, we only used two mark types (bars and circles), and we only used five encoding channels (y-position, x-position, bar length, circle color, circle area). There are more marks and channels we could use, and this larger list can be used to construct a wide array of chart types, each with their corresponding strengths and weaknesses relative to certain types of questions.

11. https://data.worldbank.org/

Before we dive into these different chart types, though, it's important that we consider two very important principles of encoding data that can be used to evaluate the quality and usefulness of the charts we see: expressiveness and effectiveness.

Expressiveness of Encodings

In his seminal 1986 work, "Automating the Design of Graphical Presentations of Relational Information," data visualization researcher Jock Mackinlay laid out the meaning and importance of expressiveness and effectiveness:[12]

> *Graphic design issues are codified with expressiveness and effectiveness criteria. Expressiveness criteria identify graphical languages that express the desired information. Effectiveness criteria identify which of these graphical languages, in a given situation, is the most effective at exploiting the capabilities of the output medium and the human visual system.*

There's a lot in that quote, so let's start by looking more closely at the first criteria: expressiveness. In *Visualization Analysis and Design*, Munzner explains, "the expressiveness principle dictates that the visual encoding should express all of, and only, the information in the dataset attributes."

But how does this principle work in practice? How can we be sure that a particular encoding accomplishes this, and why does it matter? Let's answer these questions by considering the converse: namely, let's find ways that an encoding might *fail to express* all of, and only, the information in the dataset attributes.

To explore this topic so that it really hits home, let's first recall that three of the four data scale types—ordinal, interval, and ratio—have an inherent order to them. With these three scale types, some values are naturally greater than others, and some are less than others. Only the

12. https://research.tableau.com/sites/default/files/p110-mackinlay.pdf

nominal scale type is "unordered" in that the levels of the category don't have any inherent order to them, like gender identity, country, or hair color, because the notions of "greater than" and "less than" don't apply to levels of a nominal scale:

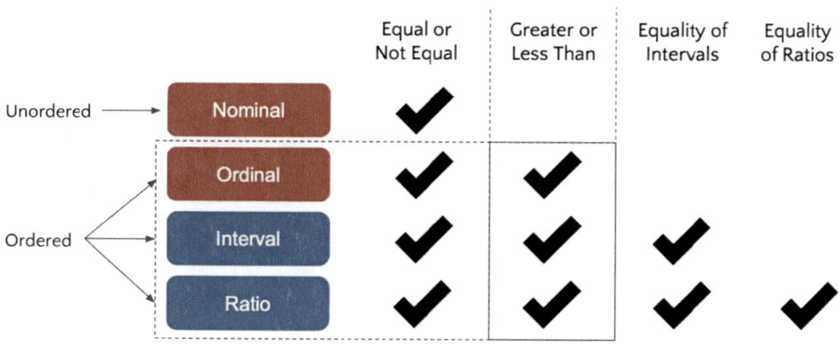

Figure 1.11. Only ordinal, interval and ratio scales are ordered

With this important fact in mind, we can find many different ways that a data visual can fail the expressiveness criteria. Let's consider three of them here, and others later in the book:

1. By showing ordered data in a way that doesn't express any order at all

Let's return to our three-country example from the previous section on marks and channels. The data field "Income Group" is ordinal, because the levels of the scale have an order to them. The order to the levels, in descending order of income, is high, upper middle, lower middle, and then finally low. Figure 1.12 shows two different ways of using the color channel to encode the levels of this variable. Figure 1.12a uses a single color hue, green, which increases in saturation as the levels of the income group increase. Figure 1.12b uses totally different colors to encode the different levels.

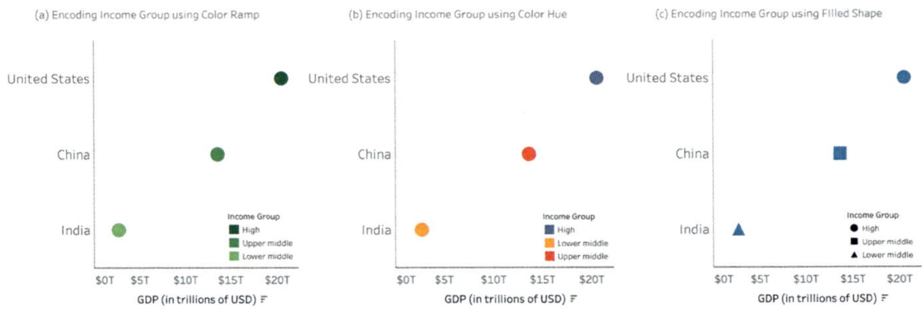

Figure 1.12. Three ways of encoding income group: (b) and (c) violate expressiveness

If we use a sequential (progressively more saturated) colormap to encode the levels of income group, as in Figure 1.12a, then the person looking at the chart with this encoding will be able to tell right away just by looking at the colors whether a certain country is in a higher income group than another. The successively increasing saturation of the green color encoding expresses this information contained in the data.

If, on the other hand, we use totally *different* colors, called hues, to encode the income group levels as shown in Figure 1.12b, then the encoding wouldn't express this ordinal information contained in the data. One is blue, one is red, and the other is orange. These colors don't express which country is in a higher or lower income group than the others. In this way, using color hue to encode an ordinal variable fails the expressiveness criteria, because it doesn't express all of the information in the data, namely the order of the income groups.

Similarly, using a different shape for each income group, as in Figure 1.12c, we can see the difference—one is a circle, another is a square, and the third is a triangle—but these shapes don't express the order to these levels either.

Of course, as mentioned before, there are always exceptions to this and every rule. For example, if we want to encode the type of medals received in the Olympic Games, then using the color hue of the medal itself would be appropriate, so long as the audience is familiar with the inherent ordering of medal type: gold > silver > bronze.

Figure 1.13. Medals are an exception to the rule that color hues don't express order

This is a special case since the names of the medals *are* color hues, and the audience might likely recognize the inherent order of these colors if they are familiar with the Olympic Games. This is what is sometimes called a "natural mapping" or a "natural encoding"—that is, an encoding that in some way reflects the actual traits of the objects or phenomena being visualized.

2. By showing ordered data in a way that expresses the wrong order

What if the encoding does express order, but it expresses the *wrong order*? For example, what if the sequential colormap used to encode the Income Group variable used a lighter green for the upper-middle-income level than the one used for lower-middle-income level, as shown in Figure 1.14?

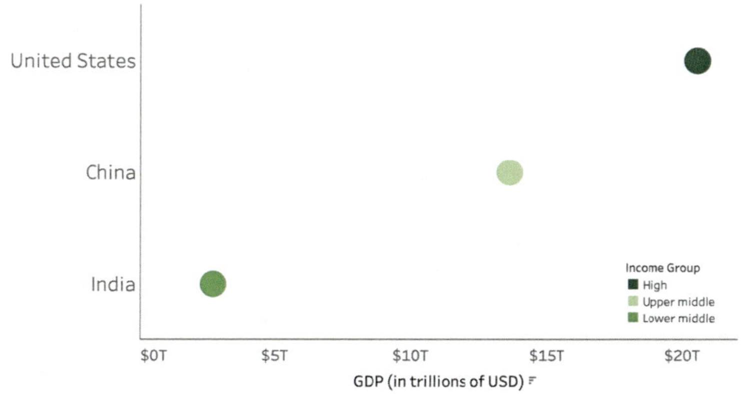

Figure 1.14. A sequential colormap that encodes the levels in the wrong order

This choice also violates the expressiveness principle because it expresses information not contained in the data. We could make the same case for using the area (2D size) channel for Income Group, as shown in Figure 1.15. If we were to use this channel, we would need to assign sizes for the levels of Income Group that were in the correct order. Figure 1.15a shows an encoding that expresses the correct order and Figure 1.15b shows an encoding that violates the expressiveness principle because the sizes are out of order.

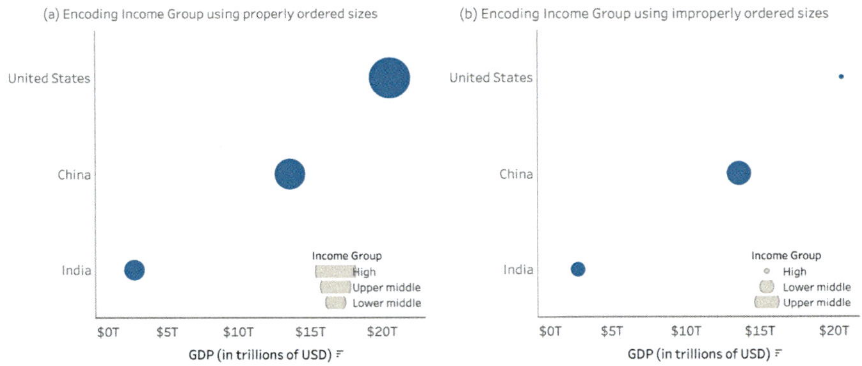

Figure 1.15. Encoding Income Group with area (2Dsize) (a) properly and (b) improperly

3. By showing unordered data in a way that expresses some order

A third way that an encoding can violate the expressiveness principle is by showing unordered data in a way that gives the impression that there is, in fact, an inherent order to the data. We know, for example, that Country is a nominal scale variable, and there's no inherent order to the different countries. We can order countries alphabetically, or by population, or a wide variety of other ways, but the countries themselves don't *naturally* belong in a specific order in the way that gold, silver, and bronze medals do.

Given this fact, any encodings that implied an inherent order to countries would be a violation of the expressiveness principle, and most likely something we would want to avoid. For example, if we consider Figure 1.16, first notice that we now have a quantitative variable, GDP (gross domestic product), determining the vertical or y-position of each mark, and another quantitative variable, population, determining the horizontal or x-position of each mark. This is a Cartesian coordinate system that allows us to compare two quantities on two axes, a convention we will consider at length in future chapters. In previous figures, each country was given its own row, so we could easily tell which mark belonged to which country. With this new encoding in Figure 1.16, we need a way to tell which of the three marks corresponds to which country.

We can see that Figure 1.16 shows three different ways of indicating which mark belongs to which country—shape, area (2D size), and a sequential colormap:

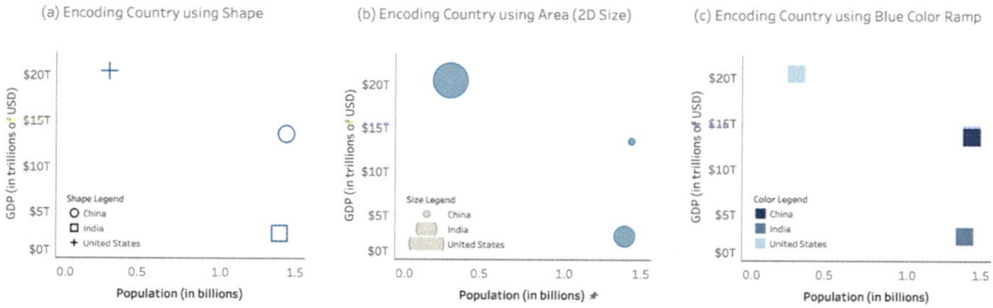

Figure 1.16. Encoding a nominal variable (Country) using three different channels

Of the three encodings shown in Figure 1.16, only Figure 1.16a upholds the expressiveness principle because the different shapes used to represent the three countries don't express any order. In Figure 1.16b, arbitrarily encoding the United States with a larger circle seems to indicate that it is in some way greater than both China and India, which are encoded using smaller circles. Similarly, in Figure 1.16c, the high saturation of the square for China seems to indicate that China is greater or contains more of something than the squares for the other two countries. In this way, the encodings in Figure 1.16b and Figure 1.16c express *more than* what the data itself expresses: an order that isn't there.

Expressiveness relates to more than just order, and we'll consider more scenarios as we go along. In general, we can think of the expressiveness principle like the paint stripes marking both sides of a lane of traffic. They keep the driver—the person creating the visual—from steering too far away from the center of the lane. But as we all know, it's quite possible to stay within the lanes and still drive poorly. The next principle, effectiveness, gives us guidance on how visual encodings can get us safely to our final destination of accurate knowledge.

Effectiveness of Encodings

Once we have established that a visual encoding follows the expressiveness principle, and shows all of, and only, the information contained in the data itself—nothing more, nothing less—then we'll need to determine whether the visual encoding is *effective* for our purposes.

But how are we to judge an encoding's effectiveness? There are many criteria we could use to judge whether the visual is effective. For example:

- Did the intended audience take note of it?
- Did they rate it highly or report a favorable experience after seeing it?
- Did they remember it some time later?
- Did it have the intended impact on them?
- Were they able to answer a question or perform a task?
- Were they able to interpret the encoding quickly and accurately?
- Some combination of the above criteria?

An argument could be made that each of these criteria, and many more, are valid ways to assess a chart's effectiveness. Ultimately, a chart is effective if it accomplishes what it was intended to accomplish, so long as the intended goal is ethical. And different situations will call for different criteria to be used. A chart published by a journalist on a news site that needs to catch the readers' attention first may have slightly different criteria of effectiveness than a chart presented to a captive audience of business leaders trying to understand a looming competitive threat. In general, it's good to embrace this messiness of the real world, and to acknowledge that there will be "different strokes for different folks."

That being said, there is some value to more narrowly defining "effectiveness" according to one or more criteria and then studying those criteria to understand them well. For example, let's consider the following pair of charts using our country population data. Would it be more effective to encode country population using the length (1D size) channel, or using the area (2D size) channel, as shown in Figure 1.17?

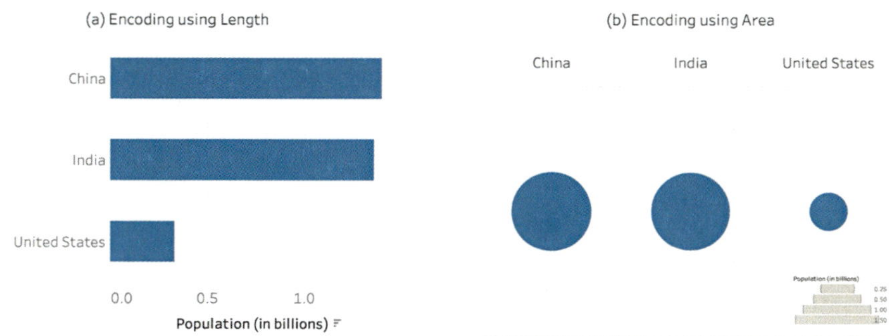

Figure 1.17. Encoding population using (a) bar length and (b) circle area

At first blush, it seems clear that the bar version of the chart is more effective than the circle version, because while it's fairly easy to see that the bar for China is slightly longer than the bar for India, it's very difficult to tell which of their two circles is larger. Can you tell which circle is larger just by looking at them? The bar length comparison is made even easier by virtue of that fact that the bars are aligned to the same baseline on the left. Due to this alignment, we're able to compare the lengths of the bars by looking at the position of their rightmost edges.

But there's another reason why the bars may be considered more effective than the circles. If we try to guess *how much longer* the mark for India is than the mark for the United States, it turns out that we tend to do a better job guessing accurately if the marks are bars than if they are circles. Try it yourself! How many of the bars for the United States do you think you could fit inside the bar for India? Is it easier or more difficult to guess how many of the circles for the United States fit inside India's circle?

We know by looking at the table of numbers in Figure 1.8 that the population of India (1.353 billion) was a little more than four times larger than the population of the United States (327 million) in 2018. To be more precise, we could fit 4.14 of the United States bars inside the

India bar, and we could fit 4.14 of the United States circles inside the India circle.

But most people have a harder time guessing accurately when they compare circle sizes than when they compare bar lengths. This fact was measured by the same Stanley Smith Stevens who gave us the four data scale types. He also gave us Stevens's Psychophysical Power Law, conceptually depicted in Figure 1.18, which shows that as the intensity of various physical stimuli are increased, our perception of the degree of increase is different for different stimuli.

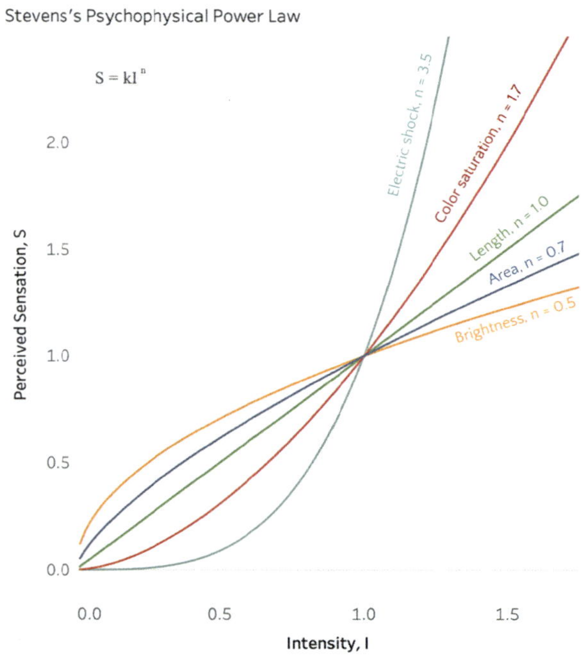

Figure 1.18. Stevens's Psychophysical Power Law[13]

Stevens found in his experiments that we tend to perceive a stimulus like electric shock as increasing more than it actually does, which explains the upward curve of the turquoise line. And we tend to perceive a stimulus like brightness as increasing less than it actually does, which explains the orange line that curves downward. Importantly, we tend to

13. https://en.wikipedia.org/wiki/Stevens%27s_power_law

perceive length as increasing in perfect proportion to its actual increase, and area as slightly less than actual, which explains the straight green line (length) and the slightly downward curving blue line (area). This graph illustrates that we tend to do a better job judging quantities accurately using length as compared to area. It's not because we're more or less intelligent than anyone else; rather, that's just how our minds work.

Much research into the effectiveness of encodings has focused on understanding more about our ability to accurately decode them as chart readers. More specifically, researchers have sought to better understand our ability to carry out perceptual tasks and form accurate judgments of quantitative information based on the encodings in the charts we see. We shouldn't fall into the trap of thinking that this is the *only* way of gauging effectiveness of a visual encoding. But it's certainly an important consideration in the vast majority of cases. In a nutshell, the question we're seeking to answer is: how accurately can the interpreter of the chart decode the encodings included in it?

Researchers William Cleveland and Robert McGill published a seminal paper on the topic in 1984, titled "Graphical Perception: Theory, Experimentation and Application to the Development of Graphical Methods."[14] This paper presented the results of tests they performed in which research participants judged proportions in data based on different encodings they were shown. In these experiments, the researchers showed participants different chart types and then they asked them to perform specific mental tasks.

To get an idea about the tasks in Cleveland and McGill's experiments, take a look at Figure 1.19, which shows an approximate re-creation of two charts included in one of their tests: a pie chart and a bar chart. Participants were asked to indicate which of the sectors of the pie chart in Figure 1.19a or which of the aligned bars in Figure 1.19b was the largest. Then they were asked "to judge what percent each of the other segments or bars is of the largest."

14. http://euclid.psych.yorku.ca/www/psy6135/papers/ClevelandMcGill1984.pdf

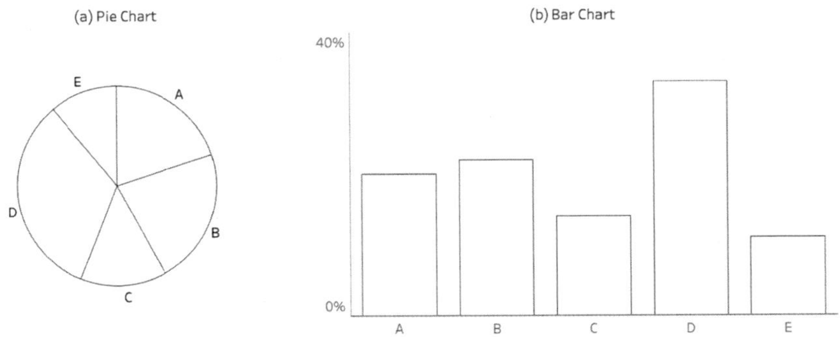

Figure 1.19. An illustrative representation of Cleveland and McGill's experiments

Try it for yourself. It's fairly easy to see that segment D is the largest in the pie chart, and bar D is the longest in the bar chart. Now consider pie segment E and bar E, for example. In each case, what percent of D is E? Is E 50%, or half of D? Is it 25%, or a quarter of D? Or is it between 25% and 50%? Do you feel more confident in the accuracy of your judgment when using the pie segments or the bars?

In both cases, the correct answer is that E is 33%, or one-third the magnitude of D. What Cleveland and McGill found was that experiment participants tended to produce more accurate judgments using the bars, and their judgments were less accurate—that is, their judgments tended to be farther away from the *actual values*—when using the pie segments.

The other pie segments are identical in proportion to the other bars as well. Test yourself using the other segments and bars and then compare your guesses to the table of actual values below:

Part	Percent of Total	Percent of Part D
A	20%	60.6%
B	22%	66.7%
C	14%	42.4%
D	33%	100.0%
E	11%	33.3%
TOTAL	100%	

Figure 1.20. The table of figures used to create the pie chart and bar chart in Figure 1.16.

Cleveland and McGill conducted further experiments, and published a second paper in 1985 titled "Graphical Perception and Graphical Methods for Analyzing Scientific Data."[15] So what did Cleveland and McGill conclude from their various experiments? They felt that they could order "elementary tasks by accuracy" from "most accurate to least" as follows:

1. Position along a common scale
2. Position on identical but nonaligned scales
3. Length
4. Angle, Slope
5. Area
6. Volume, Density, Color saturation
7. Color hue

Cleveland and McGill summarized their perspective about what makes a chart effective in the introduction to their second research paper, as follows:

> *When a graph is constructed, quantitative and categorical information is encoded, chiefly through position, shape, size, symbols, and color. When a person looks at a graph, the information is visually decoded by the person's visual system. A graphical method is successful only if the decoding is effective. No matter how clever and how technologically impressive the encoding, it fails if the decoding process fails.*

More recently, in 2010, researchers Jeffrey Heer and Mike Bostock replicated and expanded upon Cleveland and McGill's work in a paper titled "Crowdsourcing Graphical Perception: Using Mechanical Turk to Assess Visualization Design".[16] These and other experiments have been summarized elegantly by Tamara Munzner in Figure 1.21. In this diagram, Munzner places encoding channels that tend to result in higher accuracy judgments at the top of the list, and ones that tend to involve less accurate judgments toward the bottom of the list: notice the vertical "Effectiveness" axis in the center of the diagram.

15. https://www.jstor.org/stable/1695272
16. http://vis.stanford.edu/files/2010-MTurk-CHI.pdf

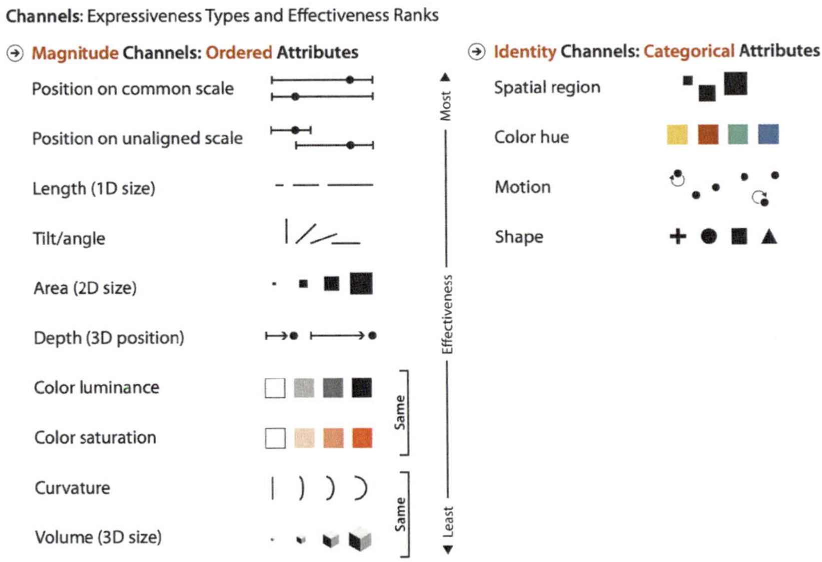

Figure 1.21. Munzner's summary diagram of the effectiveness of encoding channels

This diagram also helps us quickly assess whether an encoding violates the expressiveness principle. You will want to earmark or bookmark this page, because we will be referring back to this diagram often.

Notice that there are two sides to Munzner's diagram. The left side of the diagram applies to "Ordered Attributes," or the ordinal, interval, and ratio scales which all have an inherent order to their levels. The channels on the left side of the diagram are all "Magnitude Channels," meaning they are suitable for encoding quantities. The right side of the diagram applies to categories without inherent order to them, what Stevens called the nominal scale. The channels on the right side of the diagram are all "Identity Channels," and they are suitable for encoding unordered categories.

For example, if you go back and look at the examples of encodings that violated the expressiveness principle in the previous section, you'll notice that they all have one thing in common: they use the channels on the wrong side of Munzner's diagram as encodings:

- Income Group is an ordinal (ordered) variable (belongs on the left side of the diagram), so if a chart uses a channel on the right side of the diagram, such as color hue or shape channel to encode it, then the levels of Income Group won't appear to have any order to them. Refer back to Figure 1.12 for an illustrative example.

- Country is a nominal (unordered) variable (belongs on the right side of the diagram), so if a chart uses a channel on the left side of the diagram, such as area or sequential colormap (color saturation) to encode it, then the levels of Country will misleadingly appear to have an order to them. Refer back to Figure 1.16 for an illustrative example.

But this diagram takes us beyond expressiveness to effectiveness. The effectiveness principle is a helpful rule of thumb that states that, in general, we should use encoding channels toward the *top of the list* for variables that are the *most important to us* in a particular situation. Let's consider an example to illustrate this point. If making accurate comparisons between the GDP per capita (the gross domestic product divided by the population) of different countries was the most important task for us to carry out, then we would want to use a chart that encodes GDP per capita with a channel that's as high up on the list as possible.

Figure 1.22a would be frustrating for this purpose, because it uses the color saturation channel to encode GDP per capita. In comparison, Figure 1.22b would be much more useful because it uses the horizontal x-position channel instead. The reason for this difference in the usefulness of these two charts is that in Munzner's diagram, "Position on a common scale" is all the way at the top of the effectiveness ranking, and "Color saturation" is toward the bottom. Color saturation of the circles in Figure 1.22a lets us see quickly and easily which country has the highest GDP per capita: the United States. But as soon as we start to pose questions about the relative magnitudes of these quantities, the color saturation encoding becomes very challenging.

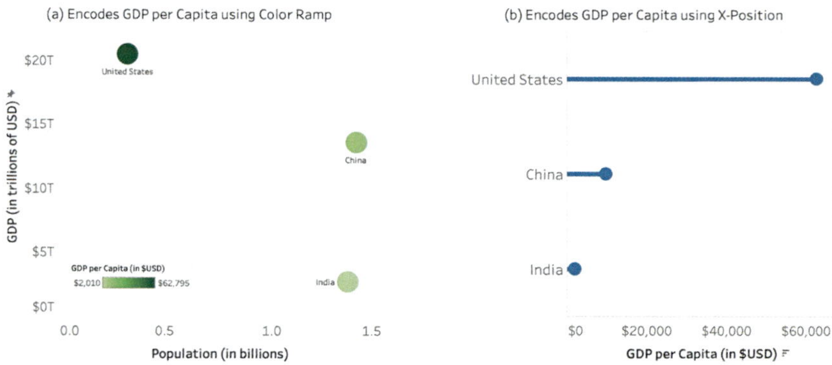

Figure 1.22. Comparing two GDP per capita encodings: (a) color saturation and (b) position

This brings up an important final point with respect to assessing the effectiveness of a chart or encoding: it depends on what's important to you. If, instead of GDP per capita, you needed to compare the size of the population along with the *overall economy* of these three countries (not per capita), then you would find Figure 1.22b to be totally useless, and you'd likely get frustrated in any attempt to make use of it. Figure 1.22a, on the other hand, would be a breath of fresh air.

Different strokes for different folks.

The Pop-Out Effect

There's another important aspect of human perception to consider when analyzing how our minds interact with visual encodings of data. Certain aspects of visuals tend to grab our attention. If the chart is designed to suit our needs, our attention will be drawn toward important aspects of the chart that we need to consider closely. On the other hand, if the chart isn't designed to suit our needs, our attention will be directed away from important aspects in our field of view, leaving us as befuddled as the audience of a magic show.

In their seminal paper "Attention and Visual Memory in Visualization and Computer Graphics," researchers Christopher G. Healey and James T. Enns introduced many of us to the concepts of "pop out" and "preattentive processing." Pop-out happens when something in the view called a "target" that's defined by a unique visual property causes it to be easily detected by a viewer—the chart reader in our case.

Consider, for example, the amount of time and effort it takes to find the red circle inside each of the three boxes of Figure 1.23:

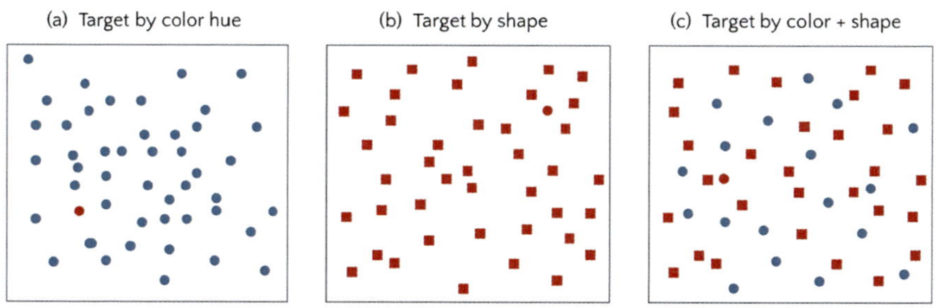

Figure 1.23. Find the red circles inside each of the three boxes

Did you notice that the task got slightly more difficult as you moved from box (a) to (b) to (c) in the Figure? The red circle among blue circles jumps right off the page or screen in Figure 1.23a. Even if the number of blue circles doubled, we'd still find the target right away. We notice the red circle before we even notice that we notice it. Our eyes are drawn to it effortlessly and almost uncontrollably. This is what's called "preattentive processing." According to Healey and Enns:

> *Typically, tasks that can be performed on large multielement displays in less than 200-250 milliseconds (msec) are considered preattentive.*

In the second box, Figure 1.23b, the red circle doesn't present itself automatically like in box (a), but we can still find the red circle among red squares incredibly quickly, and it can still be said to "pop out" of the display. The other "distractor" shapes are also red, but their straight edges make it easy to find the one with the curved shape, our red circle "target."

The task slows down considerably in Figure 1.23c, though, as the mixture of color hue and shape requires us to deliberately scan through the symbols to find the red circle. In this third box, the target does not have a single differentiating feature anymore. The target is still a red circle, but now there are other red objects: squares. And it's still a circle, but now there are also other circular objects—blue ones—that act as distractors. This doesn't make the process of finding the red circle arduous by any means, but it no longer "pops out" of the display.

Healey and Enn's paper goes on to describe other preattentive attributes that cause objects to pop out of a display in a similar way, including many of the visual encoding channels depicted in Munzner's diagram, such as the object's angle, shape, size, and luminance.

How does all of this build upon our understanding of what makes a particular chart useful and effective for us, though? Well, depending on the questions we need to answer or the tasks we need to accomplish, a data visual might help us see what we need to see, or it might make those same elements difficult to notice or find.

Consider, for example, the scenario in which a chart doesn't show just three country marks in the view, but around two hundred of them. Let's say we're still primarily interested in studying the United States, China, and India, but we'd also like to have a perspective of how these three countries compare with others, too. If we compare the three versions of the same bubble chart that encodes the percent of the population that lives in an urban environment with the life expectancy of people born into each country in 2016, we can see that only (c) on the right uses the pop-out effect to draw our attention to the targets of interest to us. The other two charts make it impossible to find our three countries of interest without some sort of interaction.

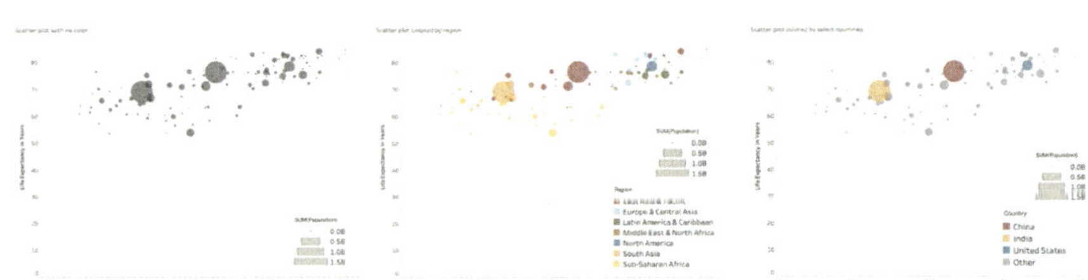

(a) (b) (c)

Figure 1.24. Using the pop-out effect to identify three specific countries

Now that's all fine and good if we happen to be interested in these three countries. But what if someone else in the audience is interested in comparing Brazil, Mexico, and Russia? If we're given the bubble chart in

Figure 1.24c, then we'd be nodding our heads at the superbly designed bubble chart while they'd be left scratching their heads wondering why they're so very unhelpful.

Designing a chart to be effective and to draw the attention of the audience to the right place isn't easy. The data can be complex, and the makeup of the audience members can be varied. Yet this makes the endeavor incredibly rich and interesting. We're seeking to combine sender, signal, and receiver in a harmonious way, so that the message that's received is both expressive and effective.

In the following lessons of this book and course, we'll consider the most commonly used chart types, and we'll learn about their most effective uses as well as ways they can sometimes mislead us. When you're done, you'll be a fluent reader of the visual language of data, something we all need to be in today's data-rich environment.

FURTHER LEARNING

"Graphical Perception: Theory, Experimentation, and Application to the Development of Graphical Methods" by William S. Cleveland and Robert McGill, 1984: https://www.jstor.org/stable/2288400

"Graphical Perception and Graphical Methods for Analyzing Scientific Data" by William S. Cleveland and Robert McGill, 1985: https://www.jstor.org/stable/1695272

"Automating the Design of Graphical Presentations of Relational Information" by Jock Mackinlay, 1986: https://research.tableau.com/sites/default/files/p110-mackinlay.pdf

"Crowdsourcing Graphical Perception: Using Mechanical Turk to Assess Visualization Design" by Jeffrey Heer and Michael Bostock, 2010: http://vis.stanford.edu/files/2010-MTurk-CHI.pdf

"Attention and Visual Memory in Visualization and Computer Graphics" by Christopher G. Healey, Senior Member, IEEE, and James T. Enns, 2012: https://www.csc2.ncsu.edu/faculty/healey/download/tvcg.12a.pdf

"Data Types, Graphical Marks, and Visual Encoding Channels" by Jeffrey Heer, 2019: https://observablehq.com/@uwdata/data-types-graphical-marks-and-visual-encoding-channels

Chapter 2

ENHANCING DATA TABLES

"Color is a power which directly influences the soul."
—Wassily Kandinsky

Individual Data Points

Solitary numbers can hold such significance for us sometimes. U.S. citizens know the exact number of years that passed between the signing of the Declaration of Independence and Abraham Lincoln's Gettysburg Address. It was four score and seven years, or 87 years, since a score represents a period of 20 years. They'll also be able to tell you the significance of the number 911, a number that often refers to the emergency response system when pronounced "nine one one," and that took on an entirely different meaning when pronounced "nine-eleven," referring to the tragic events of September 11, 2001.

Theater fans all over the world will know right away that 525,600 is the number of minutes in a common year as memorialized in the chorus of "Seasons of Love" from the Broadway musical *Rent*. They will also know that 24601 is the prison code of the character Jean Valjean in *Les Misérables* by Victor Hugo. Some will also know that Hugo supposedly chose this number for his story because he guessed that he had been conceived on June 24, 1801, or 24/6/01.

Science fiction buffs can tell you right away the temperature at which paper burns—Fahrenheit 451—or that the "Answer to the Ultimate

Question of Life, the Universe, and Everything" is the number 42 from *The Hitchhiker's Guide to the Galaxy* by Douglas Adams. And of course, what person who lived in the 1980s could forget Jenny's phone number, 867-5309?

So it is when we work with, think about, and communicate data. Sometimes we're simply looking for a single number or value—a data point. A data point all by itself lacks context and comparative power, but it can often provide us with the information we need to make a decision or to get a job done:

- What's the chance of rain this afternoon?
- How close are we to hitting our monthly sales goal?
- What percent of the vote did my favored candidate get?
- How much does the flight cost?

Of course, no data point exists in a vacuum. More than likely, as we collect these individual quantities, we are thinking about them in relation to other factors. For example, perhaps you'd like to know the chance of rain because you need to make a decision about whether or not to cancel a scheduled outdoor event. To make that decision, you'll probably need to consider other factors as well, such as the temperature, the cost of getting rained on, the cost of cancelling the event, and your tolerance for such risks.

Additionally, individual data points often beg the question—what about other, related data points? For example, as soon as you find out how close you are to hitting your monthly sales goal, you're going to want to know how much time is left, how many deals are still in the pipeline, and who is available to help you push to the goal line. Even when we think we just need a single number, data can trigger a chain of thoughts and questions that lead us down a different path than we expected.

Data points can relate to each of the four data scale types we reviewed in the previous chapter: nominal, ordinal, interval, and ratio. Let's consider once again the population data set provided by the World Bank, and consider some individual data points that might be interesting:

- Nominal scale—Country Code: COUNT of distinct values = 217
- Ordinal scale—Income Group: COUNT of low-income countries in 2019 = 29

- Interval scale—Year: MINIMUM value of year in the table = 1960
- Interval scale—Year: MAXIMUM value of year in the table = 2019
- Ratio scale—Population: Population of Canada in 1978 = 23,963,203
- Ratio scale—Population: SUM of population in 2019 = 7.6 billion
- Ratio scale—Population: AVERAGE country population in 2000 = 28,074,470
- Ratio scale—Population: MEDIAN country population in 2000 = 5,069,302

While displaying data as a simple number or figure doesn't involve any of the powerful visual encoding means we discussed in the previous chapter, any one of these data points could potentially provide us with all the information we need to get a certain job done.

Sometimes we can simply look up these values in the table itself, such as to find the population of Canada in 1978. This is an example of a single data point that might occupy an individual cell in a spreadsheet. Other times, especially for larger tables, we need to filter the table, sort it, or even run a query using SQL (Structured Query Language) against the table to find what we're looking for. Sometimes these steps will require us to perform a calculation or aggregation, such as the two queries shown in Figure 2.1 that yield two of the data points listed above.

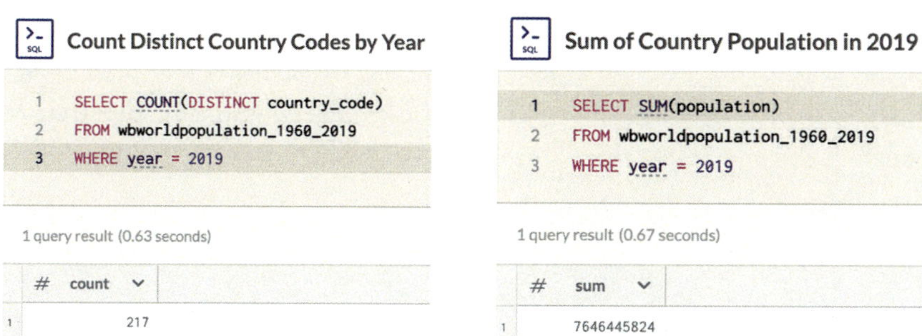

Figure 2.1. An example of two simple SQL queries that return basic aggregations

The goal of this book and course is to improve our ability to interpret the various ways of displaying data, as opposed to learning how to create them. The Data Literacy Level 2 program teaches how to create and run queries like the ones shown in Figure 2.1. Nevertheless, if you'd like to learn more and try these queries for yourself, you can access this data set as well as the query environment on the Data Literacy account on data.world.[17]

Displaying Data in Tables

We collect and gather individual data points into tables that we store for later access in spreadsheets or databases, so it makes sense that we often view the data in this same tabular form. This is nothing new. In fact, as a species we've been displaying data in tables for at least 4,500 years. Archaeologists have unearthed clay tablets dating as far back as around 2,000 BCE in Mesopotamia that have "both vertical and horizontal rulings to separate categories of information."[18] The earliest tables of data, which represent some of the first known artifacts including written language, detail transfers of sheep and goats from one ruler to another, or wages paid to temple workers, or mathematical properties like the length of sides of a right triangle.

Fast-forward to our day, and the advent of the computer along with spreadsheet software and databases mean we see tables every day. For example, Figure 2.2 shows the spreadsheet version, in MS Excel, of global population data by country available on the World Bank data site.[19]

After a couple rows of metadata (Data Source and Last Updated Date) at the top of the spreadsheet file, the table itself starts in row 4, with the column header names. If you look closely, you'll notice that the screenshot doesn't show the full table, but rather just the top left corner of the table.

17. https://data.world/data-literacy/2019-world-bank-country-population
18. https://pdfs.semanticscholar.org/ea87/b0b23e9b551fc9793b548cee3102fd5754a2.pdf
19. https://data.worldbank.org/indicator/SP.POP.TOTL

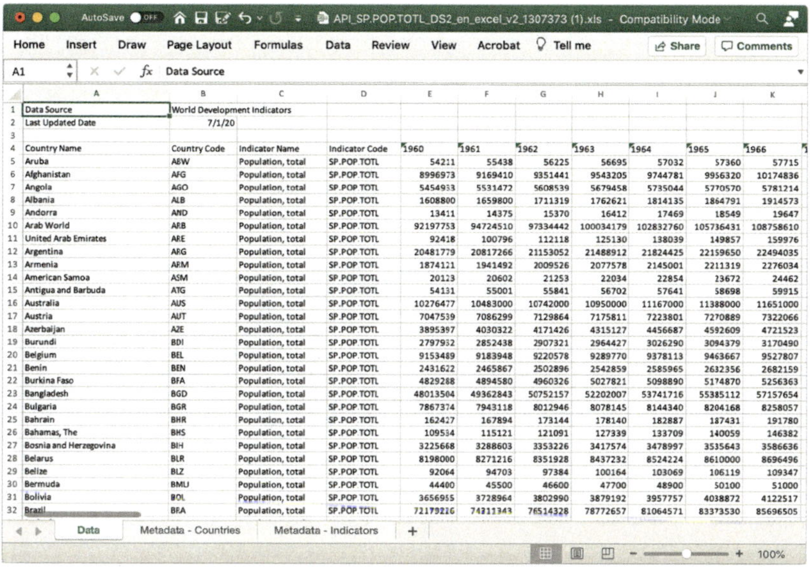

API_SP.POP.TOTL_DS2_en_excel_v2_1307373 (1).xls - Compatibility Mode

A1 Data Source

	A	B	C	D	E	F	G	H	I	J	K
1	Data Source	World Development Indicators									
2	Last Updated Date	7/1/20									
3											
4	Country Name	Country Code	Indicator Name	Indicator Code	1960	1961	1962	1963	1964	1965	1966
5	Aruba	ABW	Population, total	SP.POP.TOTL	54211	55438	56225	56695	57032	57360	57715
6	Afghanistan	AFG	Population, total	SP.POP.TOTL	8996973	9169410	9351441	9543205	9744781	9956320	10174836
7	Angola	AGO	Population, total	SP.POP.TOTL	5454933	5531472	5608539	5679458	5735044	5770570	5781214
8	Albania	ALB	Population, total	SP.POP.TOTL	1608800	1659800	1711319	1762621	1814135	1864791	1914573
9	Andorra	AND	Population, total	SP.POP.TOTL	13411	14375	15370	16412	17469	18549	19647
10	Arab World	ARB	Population, total	SP.POP.TOTL	92197753	94724510	97334442	100034179	102832760	105736431	108758610
11	United Arab Emirates	ARE	Population, total	SP.POP.TOTL	92418	100796	112118	125130	138039	149857	159976
12	Argentina	ARG	Population, total	SP.POP.TOTL	20481779	20817266	21153052	21488912	21824425	22159650	22494035
13	Armenia	ARM	Population, total	SP.POP.TOTL	1874121	1941492	2009526	2077578	2145001	2211319	2276034
14	American Samoa	ASM	Population, total	SP.POP.TOTL	20123	20602	21253	22034	22854	23672	24462
15	Antigua and Barbuda	ATG	Population, total	SP.POP.TOTL	54131	55001	55841	56702	57641	58698	59915
16	Australia	AUS	Population, total	SP.POP.TOTL	10276477	10483000	10742000	10950000	11167000	11388000	11651000
17	Austria	AUT	Population, total	SP.POP.TOTL	7047539	7086299	7129864	7175811	7223801	7270889	7322066
18	Azerbaijan	AZE	Population, total	SP.POP.TOTL	3895397	4030322	4171426	4315127	4456687	4592609	4721523
19	Burundi	BDI	Population, total	SP.POP.TOTL	2797932	2852438	2907321	2964427	3026290	3094379	3170490
20	Belgium	BEL	Population, total	SP.POP.TOTL	9153489	9183948	9220578	9289770	9378113	9463667	9527807
21	Benin	BEN	Population, total	SP.POP.TOTL	2431622	2465867	2502896	2542859	2585965	2632356	2682159
22	Burkina Faso	BFA	Population, total	SP.POP.TOTL	4829288	4894580	4960326	5027821	5098890	5174870	5256363
23	Bangladesh	BGD	Population, total	SP.POP.TOTL	48013504	49362843	50752157	52202007	53741716	55385112	57157654
24	Bulgaria	BGR	Population, total	SP.POP.TOTL	7867374	7943118	8012946	8078145	8144340	8204168	8258057
25	Bahrain	BHR	Population, total	SP.POP.TOTL	162427	167894	173144	178140	182887	187431	191780
26	Bahamas, The	BHS	Population, total	SP.POP.TOTL	109534	115121	121091	127339	133709	140059	146382
27	Bosnia and Herzegovina	BIH	Population, total	SP.POP.TOTL	3225668	3288603	3353226	3417574	3478997	3535643	3586636
28	Belarus	BLR	Population, total	SP.POP.TOTL	8198000	8271216	8351928	8437232	8524224	8610000	8696496
29	Belize	BLZ	Population, total	SP.POP.TOTL	92064	94703	97384	100164	103069	106119	109347
30	Bermuda	BMU	Population, total	SP.POP.TOTL	44400	45500	46600	47700	48900	50100	51000
31	Bolivia	BOL	Population, total	SP.POP.TOTL	3656955	3728964	3802990	3879192	3957757	4038872	4122517
32	Brazil	BRA	Population, total	SP.POP.TOTL	72179226	74311343	76514328	78772657	81064571	83373530	85696505

Data | Metadata - Countries | Metadata - Indicators

Figure 2.2. Excel version of the World Bank country population data set

Starting in column E, the country populations for each year are listed, with years increasing from left to right. The columns don't end at column K (1966). The user can scroll to the right, because the columns extend to column BL (for the year 2019) in the spreadsheet. The first 26 columns of annual data are given the letters A through Z across the top, and the 27th column is labeled AA, proceeding to AB and so on. The countries in the table are listed from top to bottom in alphabetical order, starting with Aruba in row 5 and extending down to row 268 (not shown in the view) where the population data for Zimbabwe is included in the table.

This table is pivoted, or in a "wide" format, because each of the years have their own column, stretching way out to the right. This pivoted format, with years listed side by side in adjacent columns, facilitates human comparison because we can look across columns. But we often need to unpivot such tables to a "tall" format in order to make effective use of them in today's modern analytics software and programming languages. Figure 2.3 shows these two different forms of tables and how they show the exact same information in different structures.

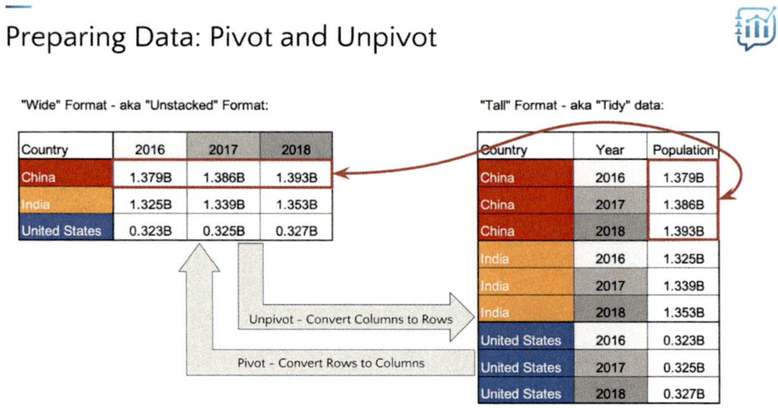

Figure 2.3. Pivoting between "tall" and "wide" structured tables

Figure 2.4 shows the unpivoted table of country population data, where all of the years can now be found in column E rather than having their own columns as in the previously shown "wide" version of the table. The table now has 13,020 rows of data instead of 264 since country population figures for different years are stacked on top of each other instead of positioned side by side.

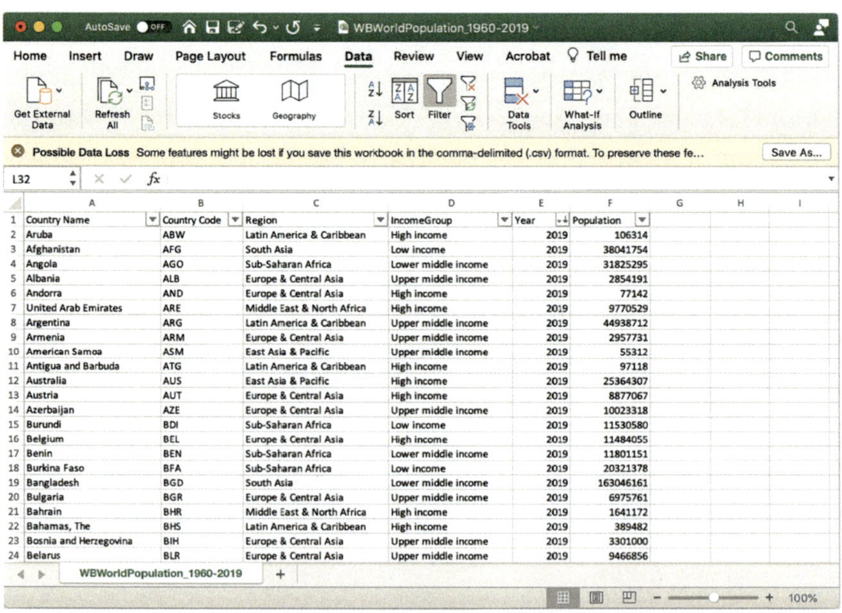

Country Name	Country Code	Region	IncomeGroup	Year	Population
Aruba	ABW	Latin America & Caribbean	High income	2019	106314
Afghanistan	AFG	South Asia	Low income	2019	38041754
Angola	AGO	Sub-Saharan Africa	Lower middle income	2019	31825295
Albania	ALB	Europe & Central Asia	Upper middle income	2019	2854191
Andorra	AND	Europe & Central Asia	High income	2019	77142
United Arab Emirates	ARE	Middle East & North Africa	High income	2019	9770529
Argentina	ARG	Latin America & Caribbean	Upper middle income	2019	44938712
Armenia	ARM	Europe & Central Asia	Upper middle income	2019	2957731
American Samoa	ASM	East Asia & Pacific	Upper middle income	2019	55312
Antigua and Barbuda	ATG	Latin America & Caribbean	High income	2019	97118
Australia	AUS	East Asia & Pacific	High income	2019	25364307
Austria	AUT	Europe & Central Asia	High income	2019	8877067
Azerbaijan	AZE	Europe & Central Asia	Upper middle income	2019	10023318
Burundi	BDI	Sub-Saharan Africa	Low income	2019	11530580
Belgium	BEL	Europe & Central Asia	High income	2019	11484055
Benin	BEN	Sub-Saharan Africa	Lower middle income	2019	11801151
Burkina Faso	BFA	Sub-Saharan Africa	Low income	2019	20321378
Bangladesh	BGD	South Asia	Lower middle income	2019	163046161
Bulgaria	BGR	Europe & Central Asia	Upper middle income	2019	6975761
Bahrain	BHR	Middle East & North Africa	High income	2019	1641172
Bahamas, The	BHS	Latin America & Caribbean	High income	2019	389482
Bosnia and Herzegovina	BIH	Europe & Central Asia	Upper middle income	2019	3301000
Belarus	BLR	Europe & Central Asia	Upper middle income	2019	9466856

Figure 2.4. The same population data "unpivoted" to list all of the years in a single column

We won't belabor the description of the spreadsheet in its various forms, though, because the fact is that we just aren't very good at using raw data tables by themselves to extract information and knowledge from data. It can be an overwhelming experience for us, especially as the number of rows and columns increase beyond just a few.

For example, you may be able to use the table in either tall or wide format to find the population of a specific country and a specific year—say, Sweden in 2017 (10,057,698)—but can you use it tell if there are any trends, patterns, or outliers within the population data? Not likely.

One way we try to simplify our lives when working with data like this is to create smaller summary tables that group or aggregate the data by categories, and that filter the data to show only specific attributes or dimensions of interest to us in a particular context. For example, we can take a categorical variable like Region (a nominal scale variable since the levels have no inherent order) and list the seven levels one above another to create seven rows, or eight if we count the header row at the top. We can then sum up the 2019 population values for all of the countries in each of the regions and place these values in the right-hand column of the table as shown in Figure 2.5.

Table: Population by Region, 2019

Nominal → Region | Population ← Ratio

Region	Population
East Asia & Pacific	2,317.0M
Europe & Central Asia	921.1M
Latin America & Caribbean	646.4M
Middle East & North Africa	456.7M
North America	365.9M
South Asia	1,835.8M
Sub-Saharan Africa	1,103.5M

Figure 2.5. Summarizing population by region for one year

Tables allow us to display more than one quantity for each of the categories listed. Interestingly, quantitative values in tables can come from all four of the data scale types, even the categorical ones—nominal and ordinal. For example, we can use a nominal scale in the table—Country Code—to create another quantity by counting the number of distinct codes in each region. With the updated version of the table shown in Figure 2.6, we can now see that even though the East Asia and Pacific region has the highest population, it only has the fourth-largest number of countries in it, with 37 countries.

Table: Population & Number by Region, 2019

Region	Number of Countries	Population
East Asia & Pacific	37	2,317.0M
Europe & Central Asia	58	921.1M
Latin America & Caribbean	42	646.4M
Middle East & North Africa	21	456.7M
North America	3	365.9M
South Asia	8	1,835.8M
Sub-Saharan Africa	48	1,103.5M

Figure 2.6. Using tables to show more than one aggregation at a time

We can take these kinds of comparisons further by using tables to view calculations not included in the data itself—ranks, averages, ratios, or percentages, for example—and we can sort the tables in different ways and add rows or columns for totals, or subtotals. All of these steps have been added in Figure 2.7, in which the regions have been sorted by total population in descending order.

Table: Population by Region, with Grand Total, 2019

Rank	Region	Number of Countries	Population	% of Total Population
1	East Asia & Pacific	37	2,317.0M	30.3%
2	South Asia	8	1,835.8M	24.0%
3	Sub-Saharan Africa	48	1,103.5M	14.4%
4	Europe & Central Asia	58	921.1M	12.0%
5	Latin America & Caribbean	42	646.4M	8.5%
6	Middle East & North Africa	21	456.7M	6.0%
7	North America	3	365.9M	4.8%
Grand Total		**217**	**7,646.4M**	**100.0%**

Figure 2.7. Using tables to add and compare calculations and totals

So far we have used the tables to break out various quantities by one single category only, namely, Region. But we know there is another categorical variable included in the data set, Income Group. We can imagine what a table that only considered Income Group would look like. We would simply replace the seven rows of region names with four rows, one for each Income Group. But how can we structure a table that shows us population totals, for example, for all of the possible combinations of region and income group?

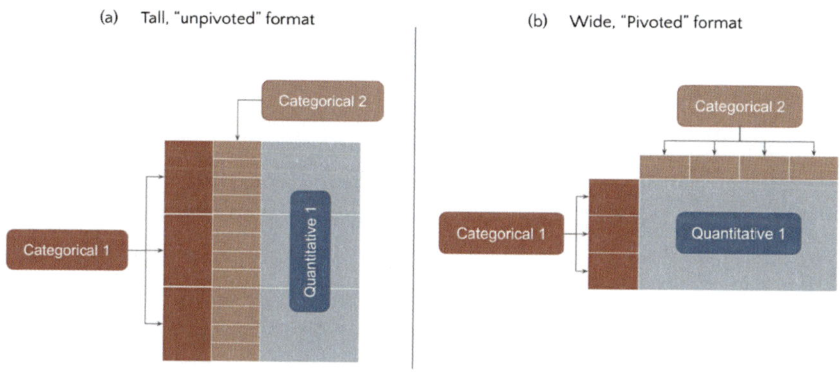

Figure 2.8. Alternative structures for 2 categorical variables and 1 quantitative variable

If we use our population data and put it into alternative (b) in Figure 2.8, then we get the table shown in Figure 2.9.

Table: Population by Region and Income Group, Sorted by Ordinality, 2019

	Income Group			
Region	Low income	Lower middle income	Upper middle income	High income
East Asia & Pacific	25.7M	296.8M	1,771.2M	223.4M
South Asia	38.0M	1,797.2M	0.5M	
Sub-Saharan Africa	534.4M	500.8M	66.9M	1.4M
Europe & Central Asia	9.3M	87.1M	303.0M	521.7M
Latin America & Caribbean	11.3M	34.3M	568.3M	32.7M
Middle East & North Africa	46.2M	197.3M	146.0M	67.3M
North America				365.9M

Figure 2.9. Population table broken out by Regions (rows) and Income Groups (columns)

How should the Income Group category levels be arranged? If we were to sort them alphabetically from left to right, then the column for "Low income" would be between the columns for "High income" and "Upper middle income." But Income Group is an ordinal scale variable, and this would not be in the order that the categories themselves are inherently in, like ordering medals by silver and then gold and then bronze. The expressiveness principle challenges us to find an order that is in sync with the natural order of the variable itself, if possible.

We can make an argument to swap, or transpose, the rows and columns, because the levels of Income Group have names that naturally connect them with the vertical orientation: "High" and "Low." Such a wording does not strictly require a vertical orientation, but there is always a small benefit to using a "natural mapping" or "natural encoding" like this. Similarly, if we were to show a table of the political spectrum, we would expect ideologies to be arranged horizontally, with liberal (or "left wing") on the left and conservative (or "right wing") on the right, per convention.

Returning to our population data table, our transposed table with both row and column totals added can be found in Figure 2.10.

Table: Population by Income Group and Region, with Totals, 2019

	Region							
Income Group	East Asia & Pacific	South Asia	Sub-Saharan Africa	Europe & Central Asia	Latin America & Caribbean	Middle East & North Africa	North America	Grand Total
High income	223.4M		1.4M	521.7M	32.7M	67.3M	365.9M	1,212.3M
Upper middle income	1,771.2M	0.5M	66.9M	303.0M	568.3M	146.0M		2,855.9M
Lower middle income	296.8M	1,797.2M	500.8M	87.1M	34.3M	197.3M		2,913.4M
Low income	25.7M	38.0M	534.4M	9.3M	11.3M	46.2M		665.0M
Grand Total	2,317.0M	1,835.8M	1,103.5M	921.1M	646.4M	456.7M	365.9M	7,646.4M

Figure 2.10. Transposed table with rows and columns interchanged, including totals

Notice that the rows are sorted with "High income" at the top, then "Upper middle income," then "Lower middle income," and finally "Low income." Notice also that the columns are sorted by total population, with the most populous region, East Asia and Pacific, on the far left, and the least populous region, North America, on the far right.

As helpful as a thoughtfully designed table can be, it does, well, leave a lot on the table (pun intended). A great table of data is like an automobile without an engine. Sure, you could push it down the street, and it'll roll downhill just fine, but it sure is missing a grand opportunity to move us around much farther and faster.

The rest of this book and course explores various ways to use what we know about the human visual system combined with helpful principles and rules of thumb to supercharge our learning from data tables. But we'll start that journey with a very simple addition to a table: color.

From Tables to Heatmaps

We know from Munzner's diagram that sequential colormap—saturation or density—is fairly low on the visual encoding channel effectiveness ranking. Often, though, it's better than using no encoding at all. And when a group or audience is adamant about receiving the data in tabular form, it can be an effective way to introduce them to the power of visual encodings.

Basic text tables already have encodings, even without adding color. The rows and columns of the table are effectively position encodings on the right-hand side of Munzner's diagram. They map discrete levels of categories to their respective rows and columns. Refer back to Figure 2.10. Data corresponding to the countries that belong in the East Asia and Pacific region, for example, will appear in the column (horizontal position) with that name. And data for the countries that belong in the High income group will appear in the row (vertical position) with that name. So that means that the countries that belong in both the East Asia and Pacific region and the high-income group, if any, will be included in the cell that is located at the intersection of these two table elements. Simple enough.

But when we look at the data that's there, it's just a number and nothing else. In 2019, 223.4 million people lived in high-income countries in the East Asia and Pacific region. On a very rudimentary level our eyes can perhaps spot numbers with many digits, since these numbers tend to be longer than numbers with fewer digits, but this is problematic because a table might include some numbers that are shown in different orders of magnitude: billions (B), millions (M), thousands (K), and so on. So we have to be alert to these differences in the way the numbers might be shown in the table itself.

Before we see what an upgraded version of our country population table looks like when we add color, let's take a quick detour and see how heatmaps can help us interpret data.

When you look at this table of numbers, what do you see? What statements could you make about the grid? Does anything pop out at you? How difficult is it to perform simple tasks, such as counting the number of 17s in the grid?

<table>
<tr><td>46</td><td>45</td><td>46</td><td>46</td><td>47</td><td>45</td><td>47</td><td>48</td><td>42</td><td>35</td><td>34</td><td>36</td><td>43</td><td>45</td><td>44</td><td>44</td><td>46</td><td>46</td><td>46</td><td>43</td></tr>
<tr><td>48</td><td>47</td><td>48</td><td>48</td><td>48</td><td>48</td><td>36</td><td>22</td><td>20</td><td>16</td><td>15</td><td>13</td><td>13</td><td>17</td><td>32</td><td>45</td><td>46</td><td>48</td><td>46</td><td>44</td></tr>
<tr><td>50</td><td>52</td><td>51</td><td>51</td><td>50</td><td>31</td><td>23</td><td>30</td><td>25</td><td>22</td><td>19</td><td>13</td><td>10</td><td>11</td><td>09</td><td>19</td><td>49</td><td>48</td><td>48</td><td>48</td></tr>
<tr><td>51</td><td>52</td><td>52</td><td>53</td><td>33</td><td>24</td><td>34</td><td>43</td><td>36</td><td>30</td><td>20</td><td>16</td><td>11</td><td>10</td><td>08</td><td>08</td><td>20</td><td>51</td><td>50</td><td>48</td></tr>
<tr><td>52</td><td>55</td><td>55</td><td>47</td><td>24</td><td>53</td><td>66</td><td>67</td><td>64</td><td>56</td><td>44</td><td>23</td><td>14</td><td>10</td><td>08</td><td>07</td><td>07</td><td>39</td><td>51</td><td>52</td></tr>
<tr><td>54</td><td>56</td><td>56</td><td>23</td><td>43</td><td>70</td><td>73</td><td>74</td><td>72</td><td>65</td><td>56</td><td>36</td><td>20</td><td>12</td><td>07</td><td>07</td><td>06</td><td>07</td><td>53</td><td>51</td></tr>
<tr><td>55</td><td>58</td><td>47</td><td>14</td><td>51</td><td>69</td><td>72</td><td>70</td><td>71</td><td>67</td><td>55</td><td>41</td><td>28</td><td>15</td><td>11</td><td>07</td><td>06</td><td>06</td><td>47</td><td>46</td></tr>
<tr><td>58</td><td>58</td><td>27</td><td>17</td><td>53</td><td>65</td><td>68</td><td>68</td><td>69</td><td>63</td><td>54</td><td>49</td><td>41</td><td>26</td><td>08</td><td>07</td><td>06</td><td>05</td><td>17</td><td>35</td></tr>
<tr><td>63</td><td>53</td><td>18</td><td>18</td><td>57</td><td>63</td><td>67</td><td>68</td><td>66</td><td>59</td><td>59</td><td>54</td><td>46</td><td>28</td><td>10</td><td>07</td><td>07</td><td>06</td><td>09</td><td>29</td></tr>
<tr><td>62</td><td>40</td><td>17</td><td>19</td><td>46</td><td>44</td><td>50</td><td>62</td><td>36</td><td>35</td><td>40</td><td>30</td><td>26</td><td>25</td><td>10</td><td>08</td><td>07</td><td>07</td><td>07</td><td>22</td></tr>
<tr><td>54</td><td>36</td><td>16</td><td>20</td><td>33</td><td>32</td><td>39</td><td>64</td><td>30</td><td>51</td><td>27</td><td>31</td><td>39</td><td>36</td><td>12</td><td>10</td><td>06</td><td>08</td><td>08</td><td>24</td></tr>
<tr><td>38</td><td>31</td><td>14</td><td>21</td><td>62</td><td>51</td><td>59</td><td>62</td><td>42</td><td>59</td><td>57</td><td>57</td><td>57</td><td>35</td><td>10</td><td>08</td><td>07</td><td>07</td><td>08</td><td>21</td></tr>
<tr><td>25</td><td>27</td><td>13</td><td>18</td><td>60</td><td>68</td><td>63</td><td>64</td><td>49</td><td>67</td><td>70</td><td>64</td><td>47</td><td>26</td><td>09</td><td>08</td><td>08</td><td>07</td><td>09</td><td>29</td></tr>
<tr><td>20</td><td>17</td><td>14</td><td>13</td><td>58</td><td>65</td><td>60</td><td>62</td><td>46</td><td>62</td><td>64</td><td>56</td><td>37</td><td>20</td><td>10</td><td>09</td><td>09</td><td>07</td><td>08</td><td>18</td></tr>
<tr><td>19</td><td>14</td><td>13</td><td>10</td><td>41</td><td>57</td><td>55</td><td>34</td><td>24</td><td>64</td><td>57</td><td>45</td><td>30</td><td>18</td><td>09</td><td>08</td><td>09</td><td>08</td><td>08</td><td>13</td></tr>
<tr><td>20</td><td>13</td><td>13</td><td>11</td><td>30</td><td>48</td><td>56</td><td>43</td><td>39</td><td>39</td><td>49</td><td>43</td><td>31</td><td>17</td><td>09</td><td>09</td><td>08</td><td>07</td><td>08</td><td>13</td></tr>
<tr><td>23</td><td>17</td><td>15</td><td>11</td><td>16</td><td>51</td><td>64</td><td>50</td><td>39</td><td>48</td><td>51</td><td>40</td><td>26</td><td>15</td><td>10</td><td>10</td><td>09</td><td>07</td><td>10</td><td>12</td></tr>
<tr><td>27</td><td>18</td><td>12</td><td>11</td><td>11</td><td>21</td><td>56</td><td>62</td><td>52</td><td>43</td><td>40</td><td>27</td><td>20</td><td>13</td><td>09</td><td>09</td><td>08</td><td>07</td><td>09</td><td>13</td></tr>
<tr><td>30</td><td>19</td><td>14</td><td>13</td><td>13</td><td>11</td><td>22</td><td>57</td><td>47</td><td>34</td><td>23</td><td>20</td><td>17</td><td>11</td><td>10</td><td>09</td><td>09</td><td>07</td><td>09</td><td>12</td></tr>
<tr><td>38</td><td>22</td><td>13</td><td>10</td><td>09</td><td>09</td><td>11</td><td>14</td><td>19</td><td>19</td><td>17</td><td>14</td><td>16</td><td>16</td><td>13</td><td>09</td><td>09</td><td>07</td><td>08</td><td>10</td></tr>
<tr><td>35</td><td>28</td><td>12</td><td>12</td><td>11</td><td>10</td><td>11</td><td>14</td><td>36</td><td>26</td><td>20</td><td>20</td><td>23</td><td>25</td><td>20</td><td>10</td><td>08</td><td>08</td><td>09</td><td>09</td></tr>
<tr><td>33</td><td>30</td><td>13</td><td>12</td><td>12</td><td>11</td><td>11</td><td>13</td><td>48</td><td>42</td><td>34</td><td>31</td><td>32</td><td>37</td><td>28</td><td>13</td><td>09</td><td>08</td><td>11</td><td>07</td></tr>
</table>

Figure 2.11. A tabular grid of numbers arranged into 22 rows and 20 columns

It's difficult to make much sense out of the numbers, and it's very time-consuming to find the answers to even basic questions, such as the number of 17s. With this simple task, you're being asked to group the two-digit numbers into one of two groups: either a number is equal to 17 or it is not equal to 17. This question relates to the identity of the number, as opposed to the relative magnitude, and so we can add color hue as an encoding to help us perform the task much faster, as shown in Figure 2.12.

46	45	46	46	47	45	47	48	42	35	34	36	43	45	44	44	46	46	46	43
48	47	48	48	48	48	36	22	20	16	15	13	13	17	32	45	46	48	46	44
50	52	51	51	50	31	23	30	25	22	19	13	10	11	09	19	49	48	48	48
51	52	52	53	33	24	34	43	36	30	20	16	11	10	08	08	20	51	50	48
52	55	55	47	24	53	66	67	64	56	44	23	14	10	08	07	07	39	51	52
54	56	56	23	43	70	73	74	72	65	56	36	20	12	07	07	06	07	53	51
55	58	47	14	51	69	72	70	71	67	55	41	28	15	11	07	06	06	47	46
58	58	27	17	53	65	68	68	69	63	54	49	41	26	08	07	06	05	17	35
63	53	18	18	57	63	67	68	66	59	59	54	46	28	10	07	07	06	09	29
62	40	17	19	46	44	50	62	36	35	40	30	26	25	10	08	07	07	07	22
54	36	16	20	33	32	39	64	30	51	27	31	39	36	12	10	06	08	08	24
38	31	14	21	62	51	59	62	42	59	57	57	57	35	10	08	07	07	08	21
25	27	13	18	60	68	63	64	49	67	70	64	47	26	09	08	08	07	09	29
20	17	14	13	58	65	60	62	46	62	64	56	37	20	10	09	09	07	08	18
19	14	13	10	41	57	55	34	24	64	57	45	30	18	09	08	09	08	08	13
20	13	13	11	30	48	56	43	39	39	49	43	31	17	09	09	08	07	08	13
23	17	15	11	16	51	64	50	39	48	51	40	26	15	10	10	09	07	10	12
27	18	12	11	11	21	56	62	52	43	40	27	20	13	09	09	08	07	09	13
30	19	14	13	13	11	22	57	47	34	23	20	17	11	10	09	09	07	09	12
38	22	13	10	09	09	11	14	19	19	17	14	16	16	13	09	09	07	08	10
35	28	12	12	11	10	11	14	36	26	20	20	23	25	20	10	08	08	09	09
33	30	13	12	12	11	11	13	48	42	34	31	32	37	28	13	09	08	11	07

Figure 2.12. The same table with color hue encoded to the nominal variable "Equal to 17 or Not"

With this version, we can quickly find out that there are nine instances of the number 17. As helpful as this color encoding is to us, it might not do justice to the true beauty contained in the table. Since the symbols in the table are quantitative, we can use sequential color saturation to encode the font color of the numbers, or even just the color of the cell themselves, removing the numbers altogether. If we do so using the "plasma" color palette, which is one of the color palettes in the Viridis family of palettes,[20] we see a low resolution of the image of Leonardo da Vinci's famous *Mona Lisa* painting appearing in the table, as shown in Figure 2.13, with the actual image shown in the bottom right corner of the figure for reference.

1	46	45	46	46	47	45	47	48	42	35	34	36	43	45	44	44	46	46	46
2	48	47	48	48	48	48	36	22	20	16	15	13	13	17	32	45	46	48	46
3	50	52	51	51	50	31	23	30	25	22	19	13	10	11	09	19	49	48	48
4	51	52	52	53	33	24	34	43	36	30	20	16	11	10	08	08	20	51	50
5	52	55	55	47	24	53	66	67	64	56	44	23	14	10	08	07	07	39	51
6	54	56	56	23	43	70	73	74	72	65	56	36	20	12	07	07	06	07	53
7	55	58	47	14	51	69	72	70	71	67	55	41	28	15	11	07	06	06	47
8	58	58	27	17	53	65	68	68	69	63	54	49	41	26	08	07	06	05	17
9	63	53	18	18	57	63	67	68	66	59	59	54	46	28	10	07	07	06	09
10	62	40	17	19	46	44	50	62	36	35	40	30	26	25	10	08	07	07	07
11	54	36	16	20	33	32	39	64	30	51	27	31	39	36	12	10	06	08	08
12	38	31	14	21	62	51	59	62	42	59	57	57	57	35	10	08	07	07	08
13	25	27	13	18	60	68	63	64	49	67	70	64	47	26	09	08	08	07	09
14	20	17	14	13	58	65	60	62	46	62	64	56	37	20	10	09	09	07	08
15	19	14	13	10	41	57	55	34	24	64	57	45	30	18	09	08	09	08	08
16	20	13	13	11	30	48	56	43	39	39	49	43	31	17	09	09	08	07	08
17	23	17	15	11	16	51	64	50	39	48	51	40	26	15	10	10	09	07	10
18	27	18	12	11	11	21	56	62	52	43	40	27	20	13	09	09	08	07	09
19	30	19	14	13	13	11	22	57	47	34	23	20	17	11	10	09	09	07	09
20	38	22	13	10	09	09	11	14	19	19	17	14	16	16	13	09	09	07	08
21	35	28	12	12	11	10	11	14	36	26	20	20	23	25	20	10	08	08	09
22	33	30	13	12	12	11	11	13	48	42	34	31	32	37	28	13	09	08	11

1	46	45	46	46	47	45	47	48	42	35	34	36	43	45	44	44	46	46	46
2	48	47	48	48	48	48	36	22	20	16	15	13	13	17	32	45	46	48	46
3	50	52	51	51	50	31	23	30	25	22	19	13	10	11	09	19	49	48	48
4	51	52	52	53	33	24	34	43	36	30	20	16	11	10	08	08	20	51	50
5	52	55	55	47	24	53	66	67	64	56	44	23	14	10	08	07	07	39	51
6	54	56	56	23	43	70	73	74	72	65	56	36	20	12	07	07	06	07	53
7	55	58	47	14	51	69	72	70	71	67	55	41	28	15	11	07	06	06	47
8	58	58	27	17	53	65	68	68	69	63	54	49	41	26	08	07	06	05	17
9	63	53	18	18	57	63	67	68	66	59	59	54	46	28	10	07	07	06	09
10	62	40	17	19	46	44	50	62	36	35	40	30	26	25	10	08	07	07	07
11	54	36	16	20	33	32	39	64	30	51	27	31	39	36	12	10	06	08	08
12	38	31	14	21	62	51	59	62	42	59	57	57	57	35	10	08	07	07	08
13	25	27	13	18	60	68	63	64	49	67	70	64	47	26	09	08	08	07	09
14	20	17	14	13	58	65	60	62	46	62	64	56	37	20	10	09	09	07	08
15	19	14	13	10	41	57	55	34	24	64	57	45	30	18	09	08	09	08	08
16	20	13	13	11	30	48	56	43	39	39	49	43	31	17	09	09	08	07	08
17	23	17	15	11	16	51	64	50	39	48	51	40	26	15	10	10	09	07	10
18	27	18	12	11	11	21	56	62	52	43	40	27	20	13	09	09	08	07	09
19	30	19	14	13	13	11	22	57	47	34	23	20	17	11	10	09	09	07	09
20	38	22	13	10	09	09	11	14	19	19	17	14	16	16	13	09	09	07	08
21	35	28	12	12	11	10	11	14	36	26	20	20	23	25	20	10	08	08	09
22	33	30	13	12	12	11	11	13	48	42	34	31	32	37	28	13	09	08	11

Figure 2.13. Encoding the magnitude of the numbers using a color ramp

Adding color in this way allows our visual system to spot patterns within the data that we couldn't see at all when we were looking at the raw numbers themselves. Now, we won't expect works of art to emerge from our tables when we add color to them, but we can see that doing so does allow us to spot not only patterns, but also trends and outliers as well.

20. https://cran.r-project.org/web/packages/viridis/vignettes/intro-to-viridis.html

To consider another example before we return to our country population data table, we'll turn to letter frequency in written language, a topic of study that has been very important to the development of the field of cryptanalysis. If we examine the percent of time that the 26 letters of the version of the Latin alphabet used in the English language appeared in the written texts of 15 different languages, leaving out the dotted and accented versions of these letters, we get the simple text table shown in Figure 2.14.[21]

Simple Table

Letter	Czech	Danish	Dutch	English	Esperanto	Finnish	French	German	Icelandic	Italian	Polish	Portuguese	Spanish	Swedish	Turkish
a	8.42%	6.03%	7.49%	8.17%	12.12%	12.22%	7.64%	6.52%	10.11%	11.75%	8.91%	14.63%	11.53%	9.38%	11.92%
b	0.82%	2.00%	1.58%	1.49%	0.98%	0.28%	0.90%	1.89%	1.04%	0.93%	1.47%	1.04%	2.22%	1.54%	2.84%
c	0.74%	0.57%	1.24%	2.78%	0.78%	0.28%	3.26%	2.73%	0.00%	4.50%	3.96%	3.88%	4.02%	1.49%	0.96%
d	3.48%	5.86%	5.93%	4.25%	3.04%	1.04%	3.67%	5.08%	1.58%	3.74%	3.25%	4.99%	5.01%	4.70%	4.71%
e	7.56%	15.45%	18.91%	12.70%	9.00%	7.97%	14.72%	16.40%	6.42%	11.79%	7.66%	12.57%	12.18%	10.15%	8.91%
f	0.08%	2.41%	0.81%	2.23%	1.04%	0.19%	1.07%	1.66%	3.01%	1.15%	0.30%	1.02%	0.69%	2.03%	0.46%
g	0.09%	4.08%	3.40%	2.02%	1.17%	0.39%	0.87%	3.01%	4.24%	1.64%	1.42%	1.30%	1.77%	2.86%	1.25%
h	1.36%	1.62%	2.38%	6.09%	0.38%	1.85%	0.74%	4.58%	1.87%	0.64%	1.08%	0.78%	0.70%	2.09%	1.21%
i	6.07%	6.00%	6.50%	6.97%	10.01%	10.82%	7.53%	6.55%	7.58%	10.14%	8.21%	6.19%	6.25%	5.82%	8.60%
j	1.43%	0.73%	1.46%	0.15%	3.50%	2.04%	0.61%	0.27%	1.14%	0.01%	2.28%	0.40%	0.49%	0.61%	0.03%
k	2.89%	3.40%	2.25%	0.77%	4.16%	4.97%	0.07%	1.42%	3.31%	0.01%	3.51%	0.02%	0.01%	3.14%	4.68%
l	3.80%	5.23%	3.57%	4.03%	6.10%	5.76%	5.46%	3.44%	4.53%	6.51%	2.10%	2.78%	4.97%	5.28%	5.92%
m	2.45%	3.24%	2.21%	2.41%	2.99%	3.20%	2.97%	2.53%	4.04%	2.51%	2.80%	4.74%	3.16%	3.47%	3.75%
n	6.47%	7.24%	10.03%	6.75%	7.96%	8.83%	7.10%	9.78%	7.71%	6.88%	5.52%	4.45%	6.71%	8.54%	7.49%
o	6.70%	4.64%	6.06%	7.51%	8.78%	5.61%	5.80%	2.59%	2.17%	9.83%	7.75%	9.74%	8.68%	4.48%	2.48%
p	1.91%	1.76%	1.57%	1.93%	2.76%	1.84%	2.52%	0.67%	0.79%	3.06%	3.13%	2.52%	2.51%	1.84%	0.89%
q	0.00%	0.01%	0.01%	0.10%	0.00%	0.01%	1.36%	0.02%	0.00%	0.51%	0.14%	1.20%	0.88%	0.02%	0.00%
r	4.80%	8.96%	6.41%	5.99%	5.91%	2.87%	6.69%	7.00%	8.58%	6.37%	4.69%	6.53%	6.87%	8.43%	6.72%
s	5.21%	5.81%	3.73%	6.33%	6.09%	7.86%	7.95%	7.27%	5.63%	4.98%	4.32%	6.81%	7.98%	6.59%	3.01%
t	5.73%	6.86%	6.79%	9.06%	5.28%	8.75%	7.24%	6.15%	4.95%	5.62%	3.98%	4.34%	4.63%	7.69%	3.31%
u	2.16%	1.98%	1.99%	2.76%	3.18%	5.01%	6.31%	4.17%	4.56%	3.01%	2.50%	3.64%	2.93%	1.92%	3.24%
v	5.34%	2.33%	2.85%	0.98%	1.90%	2.25%	1.84%	0.85%	2.44%	2.10%	0.04%	1.58%	1.14%	2.42%	0.96%
w	0.02%	0.07%	1.52%	2.36%	0.00%	0.09%	0.05%	1.92%	0.00%	0.03%	4.65%	0.04%	0.02%	0.14%	0.00%
x	0.03%	0.03%	0.04%	0.15%	0.00%	0.03%	0.43%	0.03%	0.05%	0.00%	0.02%	0.25%	0.22%	0.16%	0.00%
y	1.04%	0.70%	0.04%	1.97%	0.00%	1.75%	0.13%	0.04%	0.90%	0.02%	3.76%	0.01%	1.01%	0.71%	3.34%
z	1.50%	0.03%	1.39%	0.07%	0.49%	0.05%	0.33%	1.13%	0.00%	1.18%	5.64%	0.47%	0.47%	0.07%	1.50%

Figure 2.14. A simple text table showing the frequency of letter usage by language

Again, it's very difficult to make much out of this table in a short amount of time. Take a look at the table and let your eyes wander around it. Perhaps you can spot the largest number in the table, 18.91%, which is the percent of time that written Dutch text used the letter e in the analysis.

If we encode the color of each cell with a sequential color ramp as shown in Figure 2.15, patterns and outliers start to emerge and present themselves to us with very little effort on our part. We see right away the very light row of the letter q, and the relatively dark row of the letter e. Perhaps our eyes also notice higher usages of w, y, and z in the Polish language, or lower relative usages of j and k in the Italian language.

21. https://en.wikipedia.org/wiki/Letter_frequency

Heatmap with Sequential Color Palette

Letter	Czech	Danish	Dutch	English	Esperanto	Finnish	French	German	Icelandic	Italian	Polish	Portuguese	Spanish	Swedish	Turkish
a	8.42%	6.03%	7.49%	8.167%	12.12%	12.22%	7.64%	6.52%	10.11%	11.75%	8.91%	14.63%	11.53%	9.38%	11.92%
b	0.82%	2.00%	1.58%	1.492%	0.98%	0.28%	0.90%	1.89%	1.04%	0.93%	1.47%	1.04%	2.22%	1.54%	2.84%
c	0.74%	0.57%	1.24%	2.782%	0.78%	0.28%	3.26%	2.73%	0.00%	4.50%	3.96%	3.88%	4.02%	1.49%	0.96%
d	3.48%	5.86%	5.93%	4.253%	3.04%	1.04%	3.67%	5.08%	1.58%	3.74%	3.25%	4.99%	5.01%	4.70%	4.71%
e	7.56%	15.45%	18.91%	12.702%	9.00%	7.97%	14.72%	16.40%	6.42%	11.79%	7.66%	12.57%	12.18%	10.15%	8.91%
f	0.08%	2.41%	0.81%	2.228%	1.04%	0.19%	1.07%	1.66%	3.01%	1.15%	0.30%	1.02%	0.69%	2.03%	0.46%
g	0.09%	4.08%	3.40%	2.015%	1.17%	0.39%	0.87%	3.01%	4.24%	1.64%	1.42%	1.30%	1.77%	2.86%	1.25%
h	1.36%	1.62%	2.38%	6.094%	0.38%	1.85%	0.74%	4.58%	1.87%	0.64%	1.08%	0.78%	0.70%	2.09%	1.21%
i	6.07%	6.00%	6.50%	6.966%	10.01%	10.82%	7.53%	6.55%	7.58%	10.14%	8.21%	6.19%	6.25%	5.82%	8.60%
j	1.43%	0.73%	1.46%	0.153%	3.50%	2.04%	0.61%	0.27%	1.14%	0.01%	2.28%	0.40%	0.49%	0.61%	0.03%
k	2.89%	3.40%	2.25%	0.772%	4.16%	4.97%	0.07%	1.42%	3.31%	0.01%	3.51%	0.02%	0.01%	3.14%	4.68%
l	3.80%	5.23%	3.57%	4.025%	6.10%	5.76%	5.46%	3.44%	4.53%	6.51%	2.10%	2.78%	4.97%	5.28%	5.92%
m	2.45%	3.24%	2.21%	2.406%	2.99%	3.20%	2.97%	2.53%	4.04%	2.51%	2.80%	4.74%	3.16%	3.47%	3.75%
n	6.47%	7.24%	10.03%	6.749%	7.96%	8.83%	7.10%	9.78%	7.71%	6.88%	5.52%	4.45%	6.71%	8.54%	7.49%
o	6.70%	4.64%	6.06%	7.507%	8.78%	5.61%	5.80%	2.59%	2.17%	9.83%	7.75%	9.74%	8.68%	4.48%	2.48%
p	1.91%	1.76%	1.57%	1.929%	2.76%	1.84%	2.52%	0.67%	0.79%	3.06%	3.13%	2.52%	2.51%	1.84%	0.89%
q	0.00%	0.01%	0.01%	0.095%	0.00%	0.01%	1.36%	0.02%	0.00%	0.51%	0.14%	1.20%	0.88%	0.02%	0.00%
r	4.80%	8.96%	6.41%	5.987%	5.91%	2.87%	6.69%	7.00%	8.58%	6.37%	4.69%	6.53%	6.87%	8.43%	6.72%
s	5.21%	5.81%	3.73%	6.327%	6.09%	7.86%	7.95%	7.27%	5.63%	4.98%	4.32%	6.81%	7.98%	6.59%	3.01%
t	5.73%	6.86%	6.79%	9.056%	5.28%	8.75%	7.24%	6.15%	4.95%	5.62%	3.98%	4.34%	4.63%	7.69%	3.31%
u	2.16%	1.98%	1.99%	2.758%	3.18%	5.01%	6.31%	4.17%	4.56%	3.01%	2.50%	3.64%	2.93%	1.92%	3.24%
v	5.34%	2.33%	2.85%	0.978%	1.90%	2.25%	1.84%	0.85%	2.44%	2.10%	0.04%	1.58%	1.14%	2.42%	0.96%
w	0.02%	0.07%	1.52%	2.360%	0.00%	0.09%	0.05%	1.92%	0.00%	0.03%	4.65%	0.04%	0.02%	0.14%	0.00%
x	0.03%	0.03%	0.04%	0.150%	0.00%	0.03%	0.43%	0.03%	0.05%	0.00%	0.02%	0.25%	0.22%	0.16%	0.00%
y	1.04%	0.70%	0.04%	1.974%	0.00%	1.75%	0.13%	0.04%	0.90%	0.02%	3.76%	0.01%	1.01%	0.71%	3.34%
z	1.50%	0.03%	1.39%	0.074%	0.49%	0.05%	0.33%	1.13%	0.00%	1.18%	5.64%	0.47%	0.47%	0.07%	1.50%

Figure 2.15. Encoding cell square color with color ramp determined by letter usage

We can also use a different color palette, this time the Viridis color palette itself. Doing so as in Figure 2.16 causes different insights to surface. The Dutch e is now unavoidably obvious, as is German use of the letter e and the Portuguese use of the letter a.

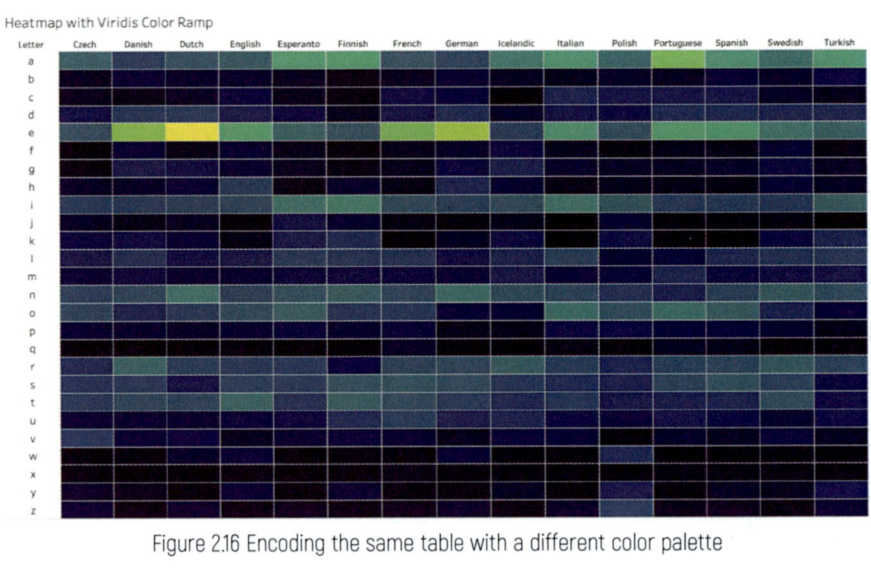

Figure 2.16 Encoding the same table with a different color palette

So we can see that the patterns we notice and the insights we glean from such encodings can depend on the details of the specific encoding we choose. Did you notice that the numbers themselves in this last version of the table were removed? This omission may make it easier to compare the colors, but there's a trade-off involved as we can't verify the precise amounts.

Design decisions often involve trade-offs like this. As we increase the density of the grid by adding more rows or columns, showing the raw numbers can become a design challenge as the cells become more and more cramped. In such cases, especially when it's critical to compare the quantities associated with two different categorical variables at the same time, heatmaps with no text can make an effective option.

If we return to our country population data table now, what happens when we encode the color of each cell with a sequential color encoding determined by the sum of the number of people who reportedly lived in each region and income group categories in 2019? We get the version of the table in Figure 2.17, with the outlier cells greater than 1.7 billion (or 1,700 million) clearly showing in the darkest saturation of the chosen blue hue.

Heatmap Table: Population by Income Group and Region, with Totals, 2019

	Region							
Income Group	East Asia & Pacific	South Asia	Sub-Saharan Africa	Europe & Central Asia	Latin America & Caribbean	Middle East & North Africa	North America	Grand Total
High income	223.4M		1.4M	521.7M	32.7M	67.3M	365.9M	**1,212.3M**
Upper middle income	1,771.2M	0.5M	66.9M	303.0M	568.3M	146.0M		**2,855.9M**
Lower middle income	296.8M	1,797.2M	500.8M	87.1M	34.3M	197.3M		**2,913.4M**
Low income	25.7M	38.0M	534.4M	9.3M	11.3M	46.2M		**665.0M**
Grand Total	**2,317.0M**	**1,835.8M**	**1,103.5M**	**921.1M**	**646.4M**	**456.7M**	**365.9M**	**7,646.4M**

Figure 2.17. The country population table converted to a heatmap

We know that color saturation is relatively low on the effectiveness ranking, so is there anything else we could do to further assist our visual cortex to make effective comparisons of the relative sizes of these numbers?

Beyond Heatmaps

If we climb slightly higher on the list of encodings on the left-hand side of Munzner's diagram, we come across Area, or 2D Size. What if we add circles inside each cell, with their areas sized proportionally to the magnitude of the numbers in each cell, allowing us to compare the populations even more effectively that way? We get the version of the heatmap shown in Figure 2.18.

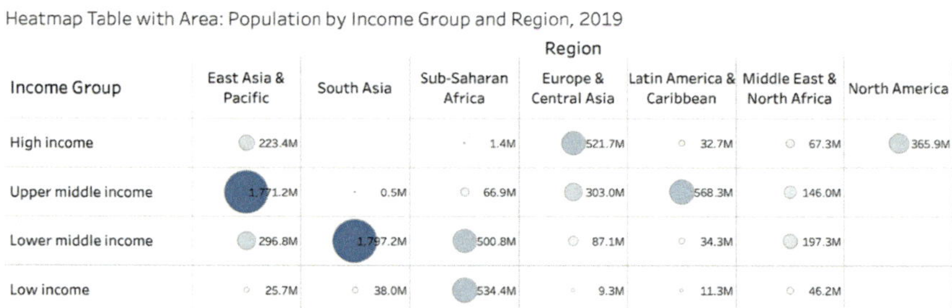

Figure 2.18. Adding the Area / 2D Size encoding to the table

As we mentioned at the beginning of this chapter, data often starts out as a table and gets converted to other forms using encodings like position, angle, size, and color. The table allows us to position individual quantities by row and column, and it affords us the opportunity to use simple encodings like color saturation to help us spot patterns and trends. In the next chapter, we'll see what happens when we make use of the most effective encodings: position and length.

FURTHER LEARNING

"The viridis color palettes" by Bob Rudis, Noam Ross and Simon Garnier, 2018: https://cran.r-project.org/web/packages/viridis/vignettes/intro-to-viridis.html

The History of Mathematical Tables: From Sumer to Spreadsheets edited by M. Campbell-Kelly, M. Croarken, R. Flood, and E. Robson, 2007: https://pdfs.semanticscholar.org/ea87/b0b23e9b551fc9793b548cee3102fd5754a2.pdf

Chapter 3

COMPARING QUANTITIES

"Accuracy of observation is the equivalent of accuracy of thinking."
—Wallace Stevens

When we consider our environment and the world around us, it's quite natural for us to size up the situation by counting things ("How many are there?") or by measuring things ("How much is there?"). Counting allows us to keep track of items using cardinal numbers (0, 1, 2, 3, and so on), and measuring involves collecting data about physical properties like weight or temperature, financial quantities like revenue or costs, temporal values like cycle time, and many more.

Immediately after counting or measuring, our minds quickly shift to categorizing and making comparisons between individuals or groups:

- Determining rank order: Which is greater than or less than another?
- Example: *"Which of two cars has a lower sticker price?"*
- Comparing magnitudes: How much greater than or less than?
- Example: *"How much less does the lower-priced car cost?"*

While such comparisons can help us understand our situation and make smarter decisions, they can sometimes be unhelpful or even harmful to us. One problem is that we can sometimes focus our comparison on a limited set of factors and ignore other relevant ones, such as comparing the sticker price of a car without taking into consideration the costs of fuel or maintenance or other factors. We may end up paying less to drive the car off the lot but far more overall before a year has passed. And that's just taking into account the financial costs of driving and not, say, the environmental ones. What we measure and compare can tell us a great deal about ourselves: our values, priorities, hopes, and fears.

This leads us to our first "Chart Reading Tip" that we'll build on to develop a checklist for ourselves by the end of the book:

TIP #1: Consider what data is shown in the display, and what relevant data might *not* be shown.

Another issue we run into when we make limited comparisons is that we tend to use them to answer totally different questions to which they simply don't apply. Our neighbor's car is newer or more expensive than ours, and so we feel inferior or somehow less of a person than they are. It sounds ridiculous when we put it like that, but these kinds of misattributions of value happen all the time. The root of the problem is that we erroneously use one metric as a proxy for another. In statistics, a proxy is defined as "a variable that is not in itself directly relevant, but that serves in place of an unobservable or immeasurable variable."[22] Sometimes proxies might be helpful, such as when we use gross domestic product as a proxy for a country's standard of living. But it's not useful to use a car value comparison to infer differences in *personal* value; the two aren't in any way correlated.

Assuming for now that the comparison we seek to make is valid and reasonable and sufficient, we must realize that we can run into further problems when we interact with visuals that don't facilitate the kind of comparisons we need to make. In other words, we need to watch out for ineffective and misleading graphical displays of data.

Munzner's diagram shown in Figure 1.20 provides us with a very useful resource to help us assess a chart and the effectiveness of its encodings. If you take a look again at the diagram, you'll note that the most effective encoding channel is *position*. It's the channel that's at the top of both

22. https://en.wikipedia.org/wiki/Proxy_(statistics)

sides of the diagram. Studies have shown that our visual system tends to be really good at using the positions of marks to make estimations about their relative magnitudes.

Ancient Counting Devices

This really shouldn't come as any surprise to us. Think back to your earliest school days when you first learned to count. Do you remember using the abacus? The position of its beads along its parallel rods allowed us to count and even carry out arithmetic functions like addition and multiplication. This ingenious device was used as an early form of calculator by people in many different places around the globe for centuries before our numeral system was even invented.[23] The Chinese suanpan and the Japanese soroban are equivalent devices dating as far back as the first century and fourteenth century, respectively.

Take a look at the sketch of the abacus shown in Figure 3.1 and see if you can figure out how the positions of the beads produce the numbers shown below each of the rods. (Hint: the positions of the beads in the top, "heavenly" section is for counting in increments of five and the positions of the beads in the bottom, "earthly" section is for counting in increments of one.)

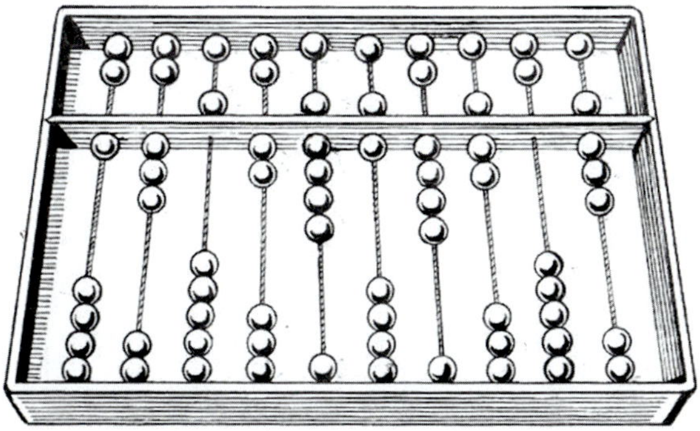

1 3 5 2 9 6 4 7 0 8

Figure 3.1. A sketch of an abacus[24]

23. https://en.wikipedia.org/wiki/Abacus
24. Placed into the public domain by its author Pearson Scott Foresman, https://commons.wikimedia.org/wiki/File:Abacus_(PSF).jpg

The abacus wasn't the only device in the ancient world that used position to enable people to keep track of quantities and to perform calculations. In South America, the Inca people used a device called the quipu to record tributary obligations, collect census records, and capture information about the calendar among other common societal tasks. The word *quipu* in their language meant "knot," and the device was made of fiber cords of cotton or alpaca hair that included knots in various places along the cords, defining segments that correspond to powers of ten.[25] Three knots in the hundreds section yielded 300, five in the tens section gave 50, and so forth. Figure 3.2 shows a sketch of an Inca quipu as conveyed in a German illustrated encyclopedia from the late nineteenth century.

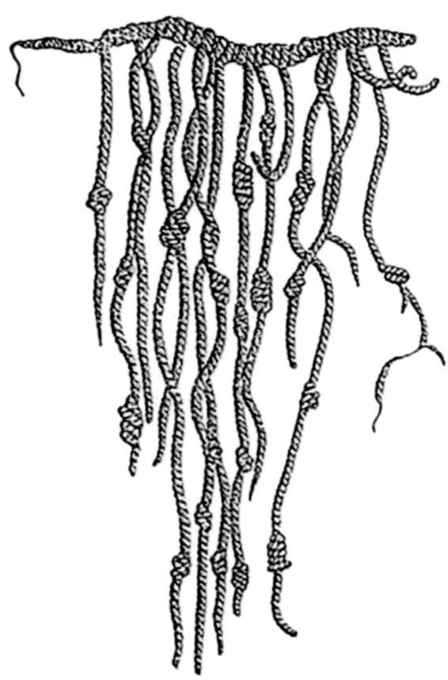

Figure 3.2. Representation of an Inca quipu[26]

These primitive but ingenious databases and accounting systems were banned by the Spanish authorities in the 1500s due to their use in sacrifices made to non-Christian gods, but hundreds of quipu can be found in museums to this day.

25. https://en.wikipedia.org/wiki/Quipu

26. Images from Meyers Konversationslexikon, 4th ed., - text figures; In the public domain: https://commons.wikimedia.org/w/index.php?curid=123557

The Dot Plot

Fast forward to today, and is it any wonder that the position encoding channel is considered by many to be the most effective way of capturing and encoding magnitude information? Modern visualization research is revealing just how ingenious those ancient devices were. Let's use the encoding channels at the top of our diagram to visualize the country population data we considered in the previous two chapters.

We can start by creating a simple **dot plot** of the number of countries in three of the regions in the world: East Asia and Pacific, South Asia, and North America. Take a look at the resulting chart in Figure 3.3. Each dot represents a unique country. Can you notice the resemblance between the dot plot and the abacus?

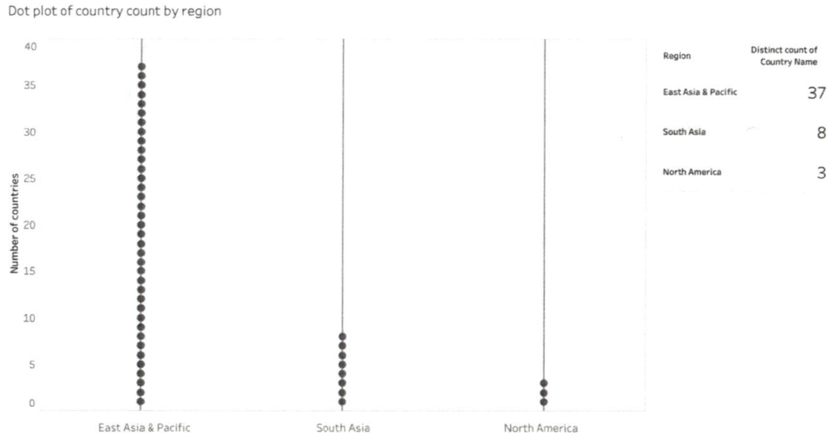

Figure 3.3. Dot plot of country count by region

The three column labels correspond to three levels of the Region variable, which is a nominal scale variable. The vertical y-axis allows us to count the countries one by one, stacked on top of each other, much like books or plates or dollar bills in the physical realm. The Country Name variable, as you'll recall, is also a nominal scale variable, but we create a new quantitative variable—Number of countries—by counting them up.

This dot plot allows us to quickly see that there are far more countries in the East Asia and Pacific region than there are in both South Asia and

North America. We can also accurately tell that there are over four times as many countries in East Asia and Pacific than in South Asia, and more than twice as many countries in South Asia as in North America. These comparisons come quickly and easily to us. In fact, it's almost impossible for us to avoid making such comparisons.

We can also visualize and compare the total populations of each of these three regions using a dot plot. We first need to compute the sum of the populations of the countries in each region, and then we can plot them. This first step of adding up the individual quantities in a group is a form of aggregation. The table in Figure 3.4 shows the updated table with the total population for each region in 2019 shown in billions of people.

Table of country count and population by region, year = 2019

Region	Distinct count of Country Name	Population
East Asia & Pacific	37	2.32B
South Asia	8	1.84B
North America	3	0.37B

Figure 3.4. Table of country count and population

The populations are very large compared to the number of countries in each region. For this reason, it would be challenging to include a dot for every single person in our new population dot plot. We can either use one dot to represent many millions of people, or we can simply use a single dot for each region and give it a position that encodes the total population. We can move the dots along the centerline as if they were beads being moved along the rod of an ancient abacus, except the distance from the base becomes proportional to the quantity that each dot encodes. Figure 3.5 shows the resulting dot plot with the vertical y-axis acting as a ruler that gives us the key to decipher the code.

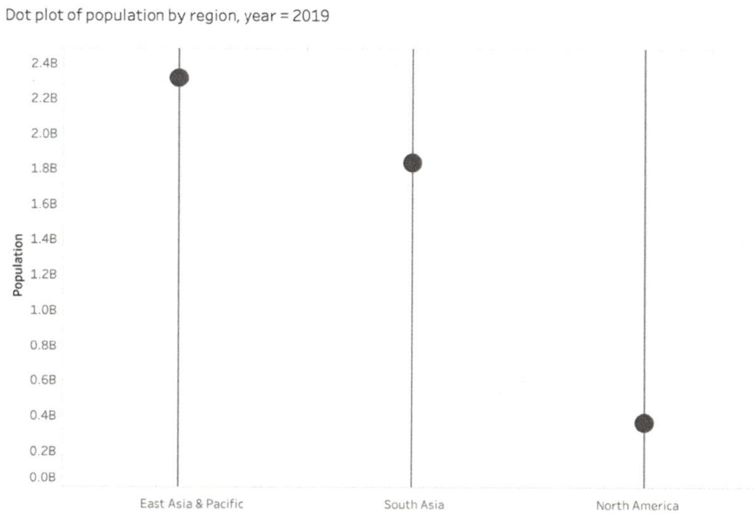

Figure 3.5. A dot plot of region population in 2019

This plot is interesting because it immediately gives us a different perspective about the relative magnitude of the different regions in 2019. While the dot plot of country count in Figure 3.3 showed us that the East Asia and Pacific region has more than four times the number of countries as the South Asia region, this second dot plot shows us that South Asia is much closer to it in population, while North America still lags quite a bit farther behind the other two regions in this metric.

The Bar Chart

At this point we can make a small stylistic shift to the dot plot that will introduce us to one of the most common single chart types in use today: the **bar chart**. This chart was introduced to us in 1786 by the founder of graphical methods of statistics, William Playfair. Playfair was an interesting character, and somewhat of a rogue. Born in Scotland in 1759, he tried his hand (and failed) at multiple careers over his lifetime. He also stormed the Bastille in 1789, he hatched a plan to devalue the French currency by creating and distributing counterfeit bills, and he even spent time in prison over the turn of the century on account of his indebtedness.

In spite of, or perhaps because of, these misadventures, Playfair managed to change the world forever by leaving us with the most common chart types in use today. It seems that he was inspired by the innovative timelines published by Joseph Priestley in 1765, just two decades prior to Playfair's first major contribution to data visualization. In Priestley's chart, shown in Figure 3.6, horizontal line lengths encode the lifespan of various influential individuals through history, giving a perspective of who predated whom, which individuals were contemporaries, and about how long they each lived:

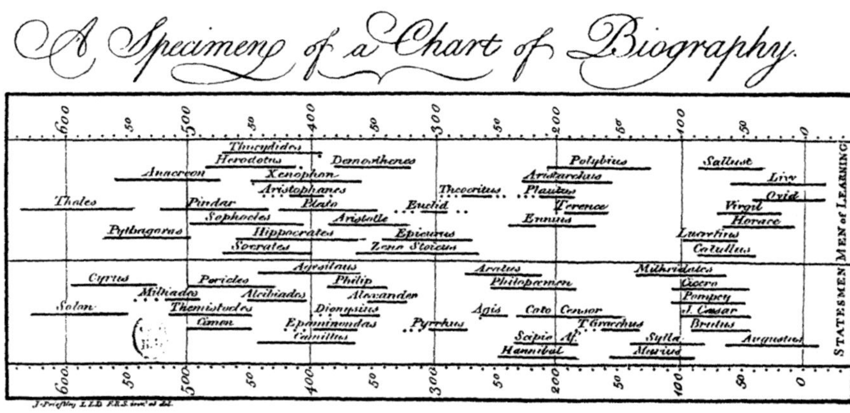

Figure 3.6. Joseph Priestley's'"A Specimen of a Chart of Biography," 1765[27]

Using this same horizontal length encoding, Playfair published his famous bar chart of Scotland's 1781 imports and exports to and from 17 different countries in his 1786 work *The Commercial and Political Atlas*, as shown in Figure 3.7.

We can read in the caption below the chart that the length of each bar is proportional to the value in British pounds of goods that Scotland either exported (the black bars) or imported (the ribbed lines) from the countries listed in the column on the far right. Thus the rows are determined by two different categorical variables: 17 different country names combined with two types of trade (import and export), resulting in 34 positions and 34 bars.

27. By Joseph Priestley (1733—1804), *Biography of William Playfair,* public domain, https://commons.wikimedia.org/w/index.php?curid=2285294

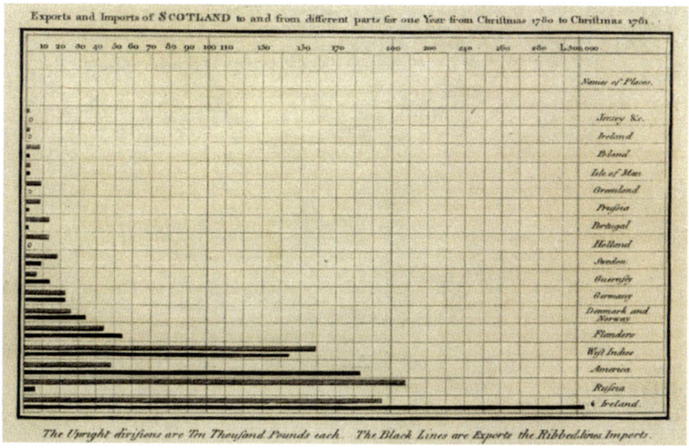

Figure 3.7. A bar chart from William Playfair's *The Commercial and Political Atlas*[28]

Following suit, we can change our abacus-like dot plot of regional population into a bar chart, albeit with vertical bars, sometimes referred to as a **column chart**. To do this, we can simply replace the dots with bars aligned at a common baseline, where population = 0, and extend each bar upward, as shown in Figure 3.8b, so that the top edges of each of the bars are located at the vertical positions that their respective dots were formerly centered. Note that this figure also includes a third version called the **lollipop chart**, which is a hybrid of the dot plot and the bar chart. In this version, the dots remain, and very thin bars connecting them to the baseline.

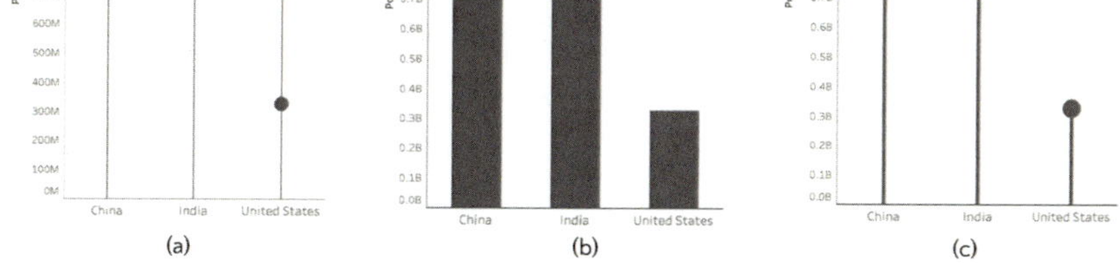

Figure 3.8. The dot plot, bar chart, and lollipop chart

28. William Playfair, *The Commercial and Political Atlas*, 1786 (3th ed. 1801); public domain.

But how do our minds decode the quantities encoded in the bar chart in Figure 3.8b? There are a few encodings we can find in it, actually, as simple as it is. Firstly, we can decode the quantities by looking at the position of the top edge of the bars, much like we judge the position of the dots in the dot plot. But also, as Cleveland and McGill pointed out in their 1984 research paper, we can also use length and area to compare the values in the bar chart:

> *But now the graphical elements used to portray the data–the bars–also change in length and area. We conjecture that the primary elementary task is judging position along a common scale, but judgments of area and length probably also play a role.*[29]

The same can be said about the lollipop chart in Figure 3.8c, though area probably plays less of a role since the stems are so thin. Notice the use of the word "conjecture" in the quote above. In Cleveland and McGill's day, and to some extent today as well, we don't have perfect and complete knowledge of the way we, as chart readers, decode different types of visual encodings. We can reason about the way visuals are constructed, and we can look at results of experiments that compare our accuracy when we use various forms of encoding. But the truth is there's a lot going on in our minds when we read and interpret visual displays of data, and we're continuing to find out more and more as researchers in the field continue to refine their experiments. If you're interested in the ongoing research, you can read about it on the Multiple Views blog on Medium.[30]

One thing we do know with a high degree of confidence, though, is that if we replace the bar chart with unaligned bars (i.e. those that don't share a common base) or shapes whose area is proportional to the quantity we're seeking to compare, as shown in Figure 3.9, then we will have a harder time guessing the true proportions accurately.

Look at both of the charts in Figure 3.9 and see if you can tell, for example, whether China or India had a larger population in 2019. It's very difficult to answer even this simple question, which is quite easy to

29. https://www.jstor.org/stable/2288400
30. https://medium.com/multiple-views-visualization-research-explained

answer using the bar chart we created earlier. And if we show the circle shapes to a group of people and then ask them to estimate how many times greater the circle for India is than the circle for the United States, their guesses will tend to be less accurate than if we ask them the same question but show them the bar chart instead. The correct answer is that the marks for India are 4.16 times as large as the marks for the United States. Which chart made this relative comparison easier, in your opinion?

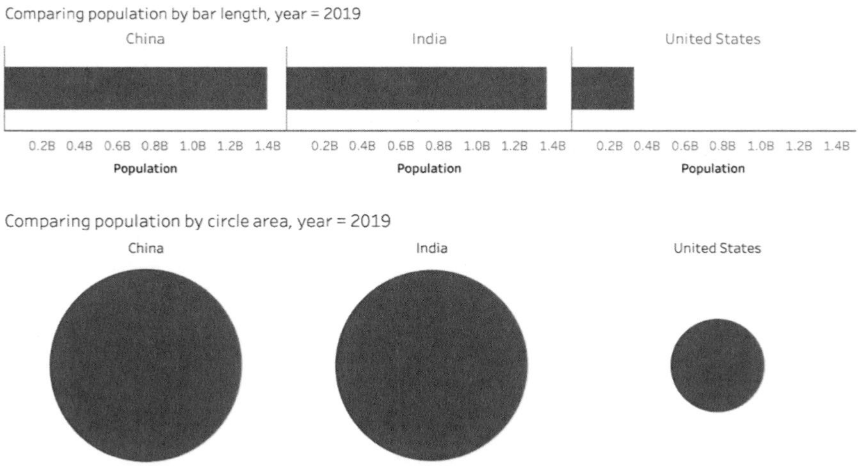

Figure 3.9. Encoding via length along an unaligned scale (top) and circle area (bottom)

What both of the encodings in Figure 3.9 make quite clear, though, is that China and India had very similar-sized populations in 2019, and that the United States had a substantially smaller one. If this is all the knowledge that is needed for a particular situation or task, then either chart would be effective enough. On the other hand, if there's any chance that the situation might require more than these basic observations, then it's likely that neither would suffice.

What about the orientation of the bars, though—should they be horizontal or vertical? We've just considered them both ways, but is there a good rule of thumb for which we should expect? Whether we prefer seeing the bars oriented horizontally or vertically can loosely depend on the topic of the data we're considering. It might be more natural to consider the topic one way versus the other.

For example, if we're comparing the heights of players on a football team, it seems like we could make a case for a vertical orientation, as we normally decide who is taller when people are standing up. Conversely, if we're comparing the length of time something takes, there's a tendency to think about time proceeding horizontally from left to right. These aren't hard-and-fast rules, but they're something to think about. The closer the chart resembles the actual phenomena it represents, the less we have to translate in our minds.

Misleading Bars

The unspoken and unwritten contract between a chart and its interpreter is that the chart won't mislead or deceive by presenting the data in a confusing way. A bar encodes its quantity in position, length, and area—the position of the starting and ending edges of the bar, the distance between the starting and ending edges, and the area of the rectangle formed by all four sides. This unstated contract between the chart and the reader would be broken if the encodings didn't faithfully convert the numbers into these graphical properties of the bars.

There are many examples of charts that mislead the reader through problematic encodings. In order to be fluent at interpreting charts, the savvy chart interpreter must be on the lookout for these types of graphical gaffes. For example, extreme cases exist where the bars aren't proportional to the data values whatsoever.[31] Instead, they might be determined by a different value in the data. Or perhaps a graphic designer just drew bars with random lengths and slapped data value labels on them. As crazy as this sounds, it happens.

Take a close look at the chart in Figure 3.10, for example. The bar lengths are actually proportional to the life expectancy, in years, of people born in each country in 2017, while the labels show the urban population in each country in the same year.

31. https://junkcharts.typepad.com/junk_charts/2020/08/testing-mess-one-chart-four-numbers-three-titles-wrong-units-wrong-data.html

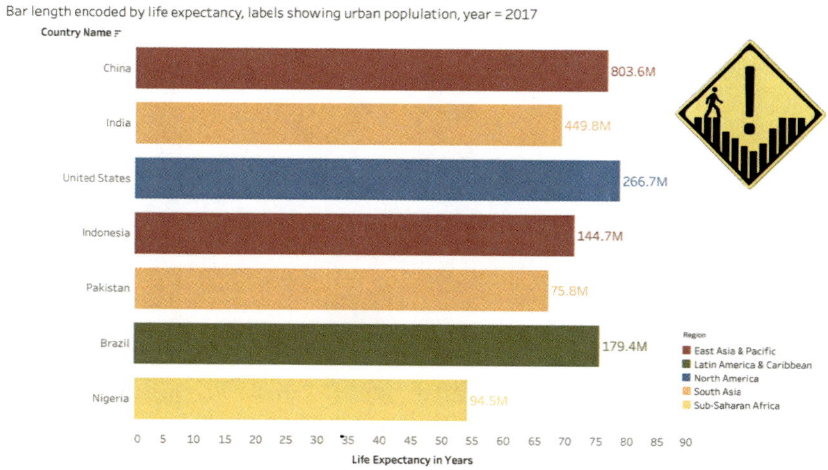

Figure 3.10. A confusing chart with labels that differ from the values used to encode bar length

This leads us to our next "Chart Reading Tip" in the list that we'll continue to build on as we go along:

> **TIP #2.** Figure out which variables in the data encode which visual channels in the chart.

A more common issue with bar charts is that the length of the bars might not be proportional to the data because the axis of the chart doesn't extend all the way to zero. For example, consider the different ways of displaying the populations of China and India in 2019, as shown in Figure 3.10.

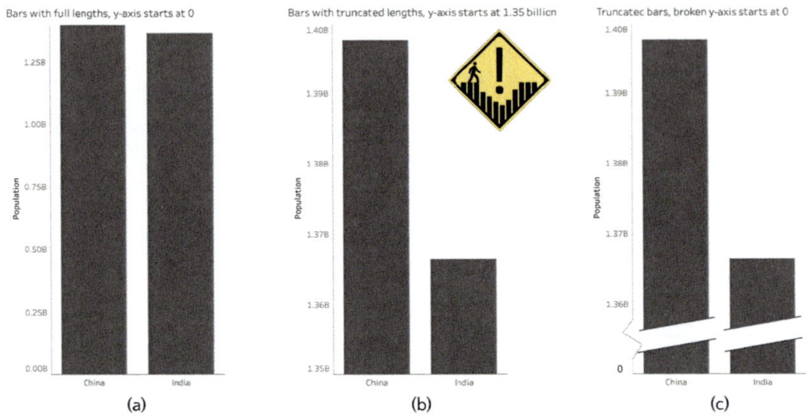

Figure 3.10. Three versions of a bar chart that differ in how they treat the continuous axis

On the left, Figure 3.10a starts the bars at a baseline corresponding to a population of zero. This makes it clear that the populations of these two countries in 2019 were almost the same, but China had a slightly larger population than India that year.

The version of the chart in the middle, Figure 3.10b, arbitrarily starts the bars at a baseline corresponding to a population of 1.35 billion; the y-axis and the bars have been truncated. This truncation makes it appear, at a glance, that the population of China was almost three times greater than the population of India in 2019. The discerning chart reader would need to pay close attention to the axis to avoid being confused in this way. The version of the chart on the right, Figure 3.10c, seeks to at least alert the reader to the truncation by breaking the y-axis and both bars and adding graphical features to indicate the break.

Some would impose an ironclad rule that the y-axis (or x-axis for horizontal bars) should *always* start at zero, with no exceptions. While this is a very good rule of thumb, there may be cases where the axis starts somewhere else, such as for values that can be negative, like profitability, or for interval scale variables that don't have an absolute zero, such as temperature. As the chart reader, we need to be aware of the continuous axes used to encode quantitative variables:

TIP #3. If there are continuous axes, notice where they start and stop.

So far we've considered dot plots and bar charts that display quantitative information for one single categorical variable at a time. First, we looked at population figures for selected regions of the world. Next, we compared the populations of a few chosen countries. But how do we use charts to compare quantities for more than one categorical variable at the same time, like we did in the previous chapter with our heatmaps, such as in Figure 2.17?

The Bar Chart Grid

Just as we were able to use heatmaps to compare quantities for more than one category, we can similarly create bar charts that include partitions for more than one categorical variable at a time. Let's say we want to quickly compare the population for countries in three different regions of the world in 2019, but we want to see each region's population broken out into the four different income groups. This would involve 12 different groups (three regions times four income groups). We can arrange the corresponding 12 bars in one of two list forms, as shown in Figure 3.11.

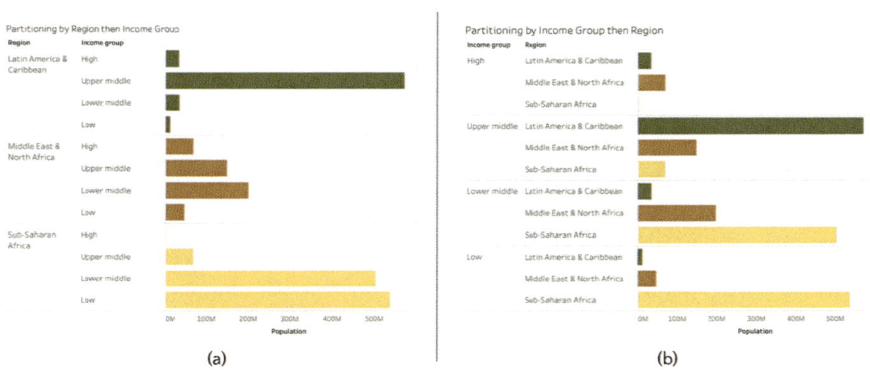

Figure 3.11. Two lists of 12 bars, switching the order of the category partitions

In Figure 3.11a on the left, we have first partitioned the data into three different regions, and then we further subdivided these three sections into four rows each, one for each income group. Notice how the income group levels are listed three times each in the second column, and that they're sorted in descending fashion according to their inherent order, with the "High" income group at the top of each partition and the "Low" income group at the bottom of each partition.

In Figure 3.11b on the right, the order of the categorical variables has been switched. The list is first partitioned into the four different income groups, again sorted ordinally, and then each of these four sections is further subdivided into three rows, one for each of the regions. Now it's the region levels that are repeated in the list.

But how are these two bar charts different, and which one is better? It depends on which comparison is more important to you. Do you want to be able to quickly and easily compare populations *within* each region? If so, the arrangement in Figure 3.11a might be more useful, as all of the bars for each region are grouped next to each other. If you're more interested in comparing within each income group, then perhaps the arrangement in Figure 3.11b would be better for a similar reason: all of the bars for each income group are located immediately adjacent to each other.

There's another way we can arrange the bars that eliminates the need to repeat any of the category levels as we had to do in both versions of Figure 3.11. We can create a 2D grid of bars, similar to the way we created the grid for the heatmap, by using one of the categorical variables to create rows and the other to create columns. For example, if we create a grid with four rows (one for each income group) and three columns (one for each of the regions we're considering), then we end up with the 4x3 bar chart grid shown in Figure 3.12.

This arrangement is appropriate because the levels of the income group category are listed vertically, with the "High" level at the top and the "Low" level on the bottom. But we could always switch the rows and columns if we wanted to, and we could also rotate the bars to be vertically oriented instead of horizontally oriented. Each of these changes would affect which bars shared a common baseline. In Figure 3.12, the bars for each region share a common vertical baseline, making it easier to compare populations of the different income groups within each region.

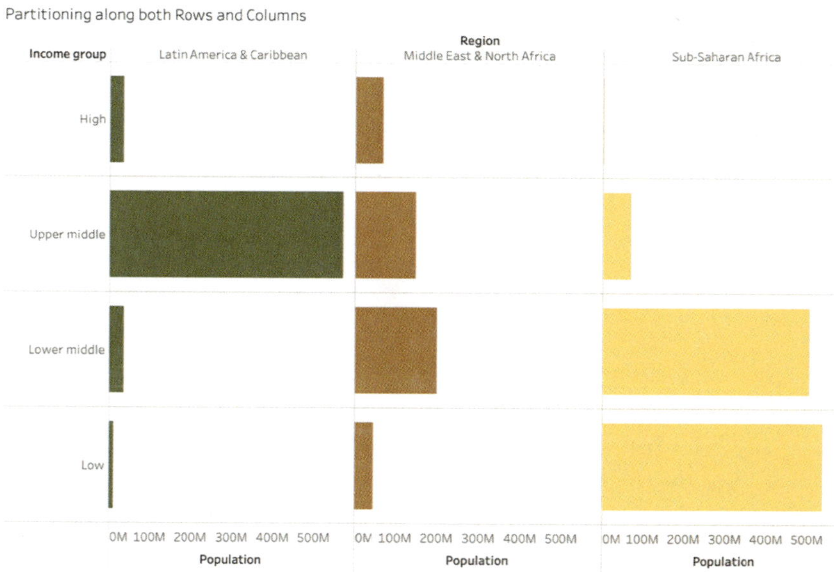

Figure 3.12. A 4x3 bar chart grid

If we keep the 4x3 grid exactly the same, but we rotate the 12 bars 90°, then we get the version of the matrix shown in Figure 3.13. Notice in this alternate version, the bars for each of the income groups now share a common horizontal baseline, making it easier to compare populations of the different regions within each income group.

What these different bar chart arrangements show us is that there are many ways to orient marks when we want to compare quantities for more than one category at a time. In order to evaluate the suitability of the chart, we need to be clear about what tasks we need to carry out, what questions we need to ask, and what comparisons would best answer those questions. At that point, we can decide whether the chart, as constructed, will help us get the job done.

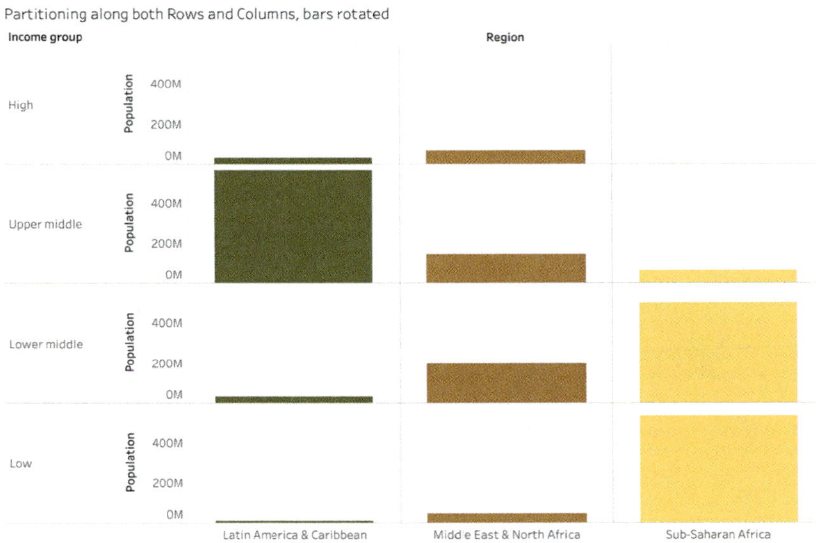

Figure 3.13. The same bar chart grid with the bars rotated 90°

This leaves us with our third tip of the chapter, and fourth overall:

TIP #4. Ask yourself whether the chart's encodings and design help you answer your most important questions.

In this chapter we considered specific countries or regions that, when put together, didn't form a complete unit. In other words, they constituted something less than one total and namable entity. But what if they do add up to a grand total that is complete or full? In the next chapter, we'll consider the encodings and chart types that help us see the part-to-whole relationships in our data.

FURTHER LEARNING

Video: "Threads That Speak: How The Inca Used Strings to Communicate" by National Geographic: https://youtu.be/AmPyz1kCbOw

"William Playfair and the Psychology of Graphs" by Ian Spence, University of Toronto: http://www2.psych.utoronto.ca/users/spence/Spence%20(2006).pdf

Chapter 4

RELATING PART-TO-WHOLE

"A whole new realism resides in the way one envisages an object or one of its parts."

–French painter Fernand Léger

There are times when we need to consider how some whole entity or amount breaks down into its various constituent parts. It might be helpful to know what percent of the total amount each part constitutes, and how these relative amounts compare to each other. Sometimes we're putting the parts together to form the whole, such as with a puzzle or a mechanical assembly. Other times we're dividing or breaking up the whole into pieces, such as with a company's annual budget. In both cases, the part-to-whole relationship is a critical consideration.

These situations aren't just commonplace in today's world; they were familiar in earlier times, too, and they prompted breakthrough innovations in mathematics. The first algebraic procedures were written down in 820 CE by a Persian polymath named Al-Khwarizmi living in Baghdad, whose Latinized name was Algorithmi, a source of our English word "algorithm." His text Al-jabr, the root of the word "algebra," gave practical advice on how to solve everyday problems related to distributing land or dividing up an inheritance in adherence with Islamic law.

For situations both ancient and modern that involve part-to-whole relationships, it would help us to know which graphical encodings express the part-to-whole relationship particularly well. And in order to become a fluent chart reader, we'll need to learn how to interpret the charts that make use of these encodings. Some common scenarios in today's world that involve part-to-whole relationships include:

- The whole: a company's monthly revenue; the parts: revenue for each region
- The whole: a company's monthly revenue; the parts: revenue for each product line
- The whole: the number of students in a class; the parts: number of students by gender
- The whole: the visits to a website one week; the parts: visits by mobile, tablet, and desktop
- The whole: all of the votes counted in an election; the parts: the number of votes for each candidate

When we're considering part-to-whole relationships, it would help to look at graphical encodings that express the part-to-whole relationship. Conversely, if we're considering something that isn't part-to-whole data, it would be helpful if the encoding doesn't give the impression that it is. This is an application of the expressiveness principle we talked about in the first chapter - namely, that the encoding should express all of, and only, the information contained in the data.

But how can the encoding give us a clue that we're dealing with components parts of a larger whole? Simply, we can make the case that some chart types convey the notion of "completeness" or "fullness" by the way the marks themselves are arranged. It will help to consider an example so that we can paint the picture.

Let's continue to build on the data set we've been using—the population of the world by country—and let's also start with the same chart type we looked at last—the bar chart. Instead of considering just a few regions, like we did at the end of the previous chapter, let's take into account the population of every single country in the world. In this way, we've identified the "whole" of the "part-to-whole" relationship.

The Stacked Bar Chart

If we start with the sum of the population of all of the 217 countries and territories listed in the World Bank population data set in 2019, we get a total population figure of about 7.646 billion people in the world that year. We can see this total value in the single bar chart in Figure 4.1a. Next to it, in Figure 4.1b, we can see the same column, but now with segments for each of the seven regions in the data set, forming a stacked bar chart.

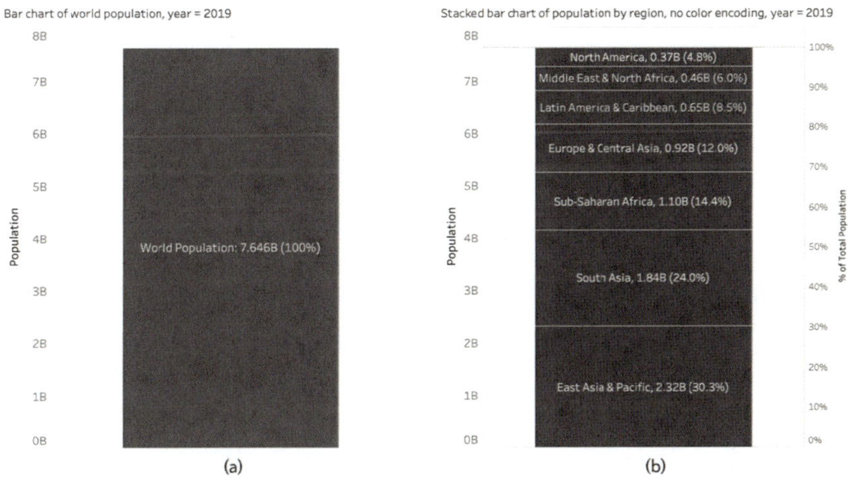

Figure 4.1. A bar chart and a stacked bar chart showing population, and by region

We can also consider two different versions of this stacked bar chart that make different use of color in Figure 4.2.

In the first revision, Figure 4.2a, the Region name has been used to encode the color hue of each of the segments. Since Region is a nominal variable without any inherent order, using color hue to encode the levels of this variable follows the expressiveness principle since the hues don't give us the impression of any order. The segments were already separated by their borders, and now they're made even more distinct by their different color hues.

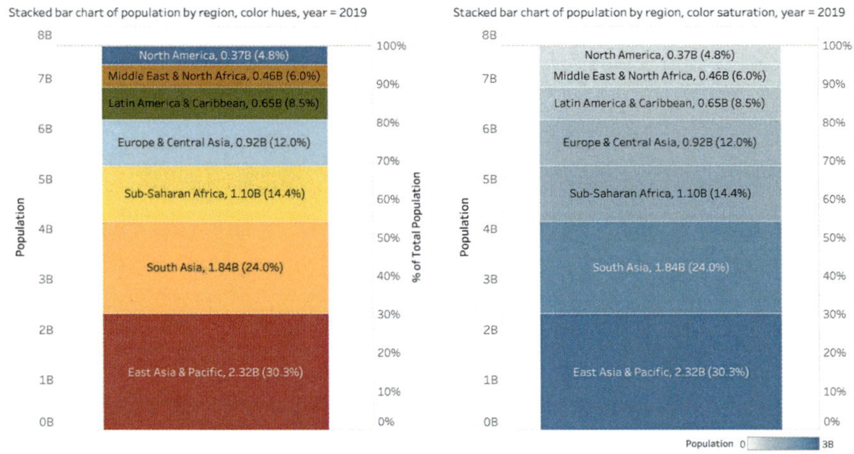

Figure 4.2. The stacked bar chart with two different uses of color

In the second revision, Figure 4.2b, the sum of each region's population has been used to encode the color saturation of each of the segments, as indicated by the legend at the bottom. Population is a quantitative variable, so using color saturation follows the expressiveness principle by giving us the impression of relative quantity. Color saturation in this example is a redundant encoding since population has already been used to encode the height of the segments, so now it's used twice in the same chart.

Notice that the stacked bar charts have two vertical y-axes; the one on the left gives us the population amount, and the one on the right gives us the percent of the total amount. In the labels of each segment, we can see the region name, the sum of the population of its countries in 2019, and the percent of the world's population accounted for by that region. For example, South Asia had a population of 1.848 billion, which made up 24.0% of the population of the world. Looking at this chart, can you tell whether the combined population of South Asia and East Asia and Pacific make up more or less than half of the world's population? They are the two segments on the bottom, and we can see that the top edge of South Asia extends just beyond the 50% mark on the right-hand y-axis.

Of course, we can always rotate the vertical bars by 90° to form a horizontal stacked bar, as shown in Figure 4.3. In the previous chapter we discussed the notion of natural encodings, but it doesn't seem like the topic of population would beg to be oriented one way or the other. We can observe, however, that the horizontal version of the chart is slightly more difficult to read because the labels inside the segments are rotated sideways in order to fit.

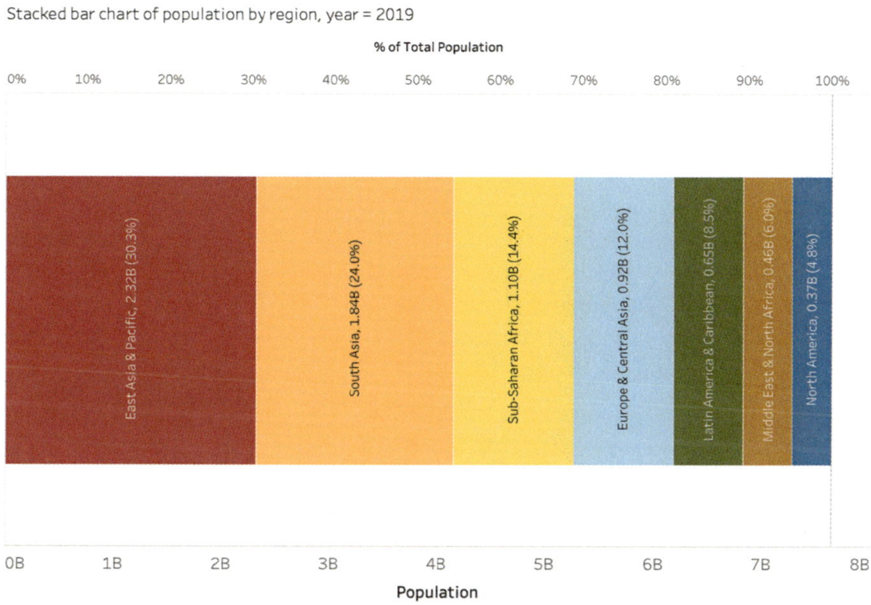

Figure 4.3. A horizontal orientation of the stacked bar chart

Another variation of the stacked bar chart is the **waterfall chart** shown in Figure 4.4b. This variation simply staggers the individual segments into their own position (either column or row), yielding a cascade effect that breaks down the total into its component parts step-by-step:

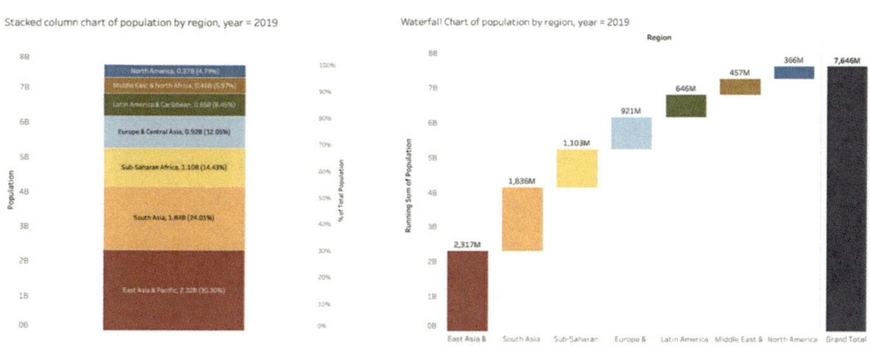

(a)
(b)

Figure 4.4. A stacked bar chart (a) and its associated waterfall chart (b)

The waterfall chart, especially when accompanied by a single bar that expresses the grand total amount, can adequately express a "part-to-whole" relationship to the chart reader. It's popular in financial scenarios, such as changes in budget by department, for example, when some segments may actually be negative. An example of a waterfall chart showing changes in budget outlay is shown in Figure 4.5, in which we see the change in allocation of federal funding from one year to the next.

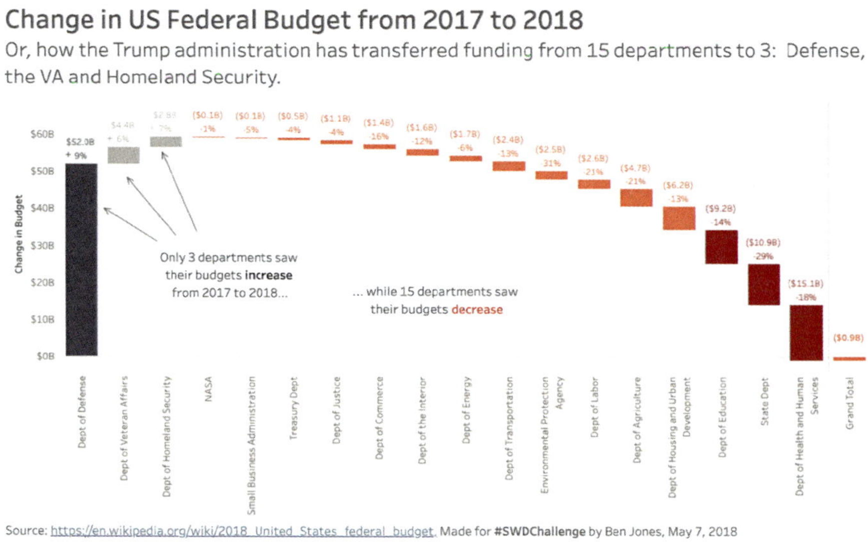

Figure 4.5. A waterfall chart showing the change in budget allocation by department

If we think about what we've learned from Munzner's diagram, we'll realize that there's a tradeoff happening with the charts we've considered so far in this chapter. Their stacked or staggered segments don't share a common baseline like the standard bar chart, so it can be slightly more difficult to accurately compare the relative sizes of each individual segment.

For cases when it's necessary to make such part-to-part comparisons with high accuracy, it might be best to use a standard bar chart, perhaps with percentage labels and a grand total bar to make it clear that you're looking at composite parts. Figure 4.6 shows an example of this type of chart.

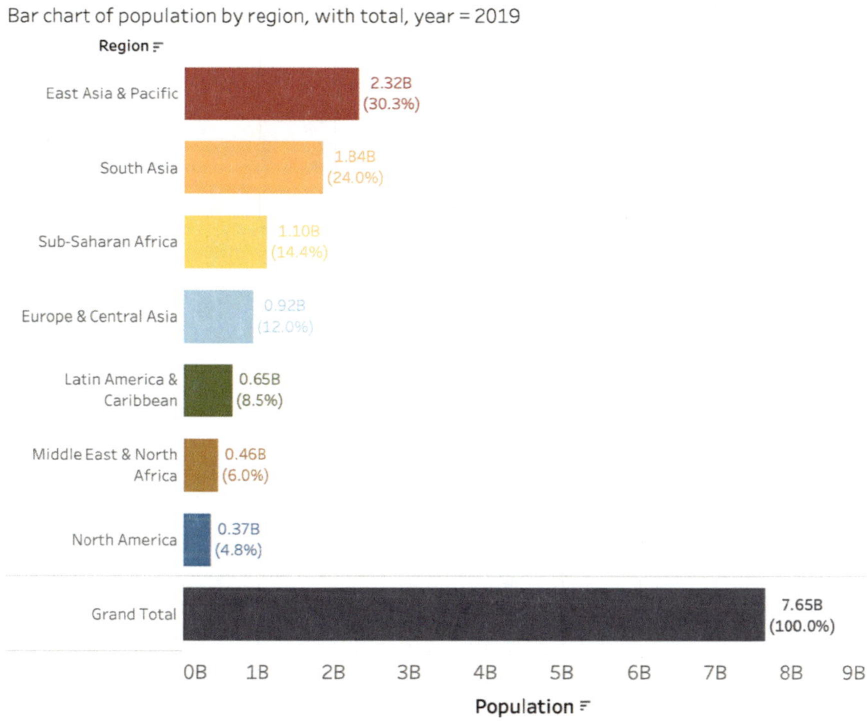

Figure 4.6. A bar chart of population by region with a total bar and percent labels

So far we've considered how to see part-to-whole information for a single categorical variable. We used our country population data set and considered the breakdown by region of the world. But how do we consider part-to-whole data when there's more than one categorical variable involved? In the previous two chapters we used partitioning to create a grid or matrix of shaded cells, circles, or bars. That proved to be an effective way to compare how quantities break down into combinations of two different categorical variables.

Instead of partitioning the view based on both of the categorical variables, we can use the first category to create multiple bars and the second to define stacked bar segments. In Figure 4.7a, we see the population of the world broken into four rows by income group, with each of the four main bars divided into seven colored segments based on region.

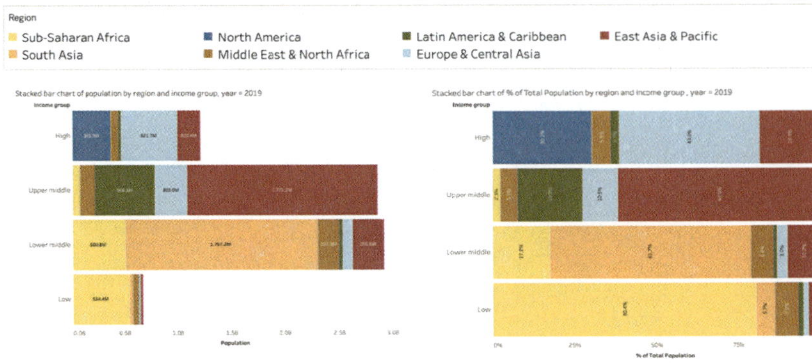

(a) (b)

Figure 4.7. A stacked bar chart of population (a) and of percent of total population (b)

On the right-hand side, in Figure 4.7b, instead of looking at the actual population values, we now see the percent of each income group's total population that live in each region. In this "percent of total" stacked bar chart, each row adds up to 100%.

So which stacked bar chart is better, the one that shows population amounts on the left or the one that shows percent of each group's population on the right? It depends on which comparisons are more important for us to make. We can use the chart on the left-hand side, Figure 4.7a, to compare actual population values, but it's difficult to use it to compare the percentages.

We can use the version on the right, Figure 4.7b, to quickly answer questions about the percent breakdown of each income group's population, but we can't use it to tell which income group had a greater overall population. In this way, neither is "better" than the other, and they can be used in tandem to provide us with a more complete perspective.

Every single one of the visual displays of data we've considered thus far has made use of a Cartesian, or rectangular, coordinate system where position and size come in the form (x,y), where x and y are perpendicular and locate a point in a 2D plane. This is a very powerful coordinate system because it's very straightforward for us to use for linear comparisons. But sometimes it can be helpful to visualize data using a polar coordinate system, where points are determined by their distance, r, as well as their angle of rotation, θ (theta) from a center point.

The Pie Chart

In this next section we'll be considering what is perhaps the most controversial of all chart types: the **pie chart**. A pie chart is simply a stacked bar chart created using polar instead of rectangular coordinates. Instead of dividing a bar into linear segments, we divide a circle in sectors. Figure 4.8 shows the population stacked bar chart on the left (a) and the corresponding pie chart on the right (b).

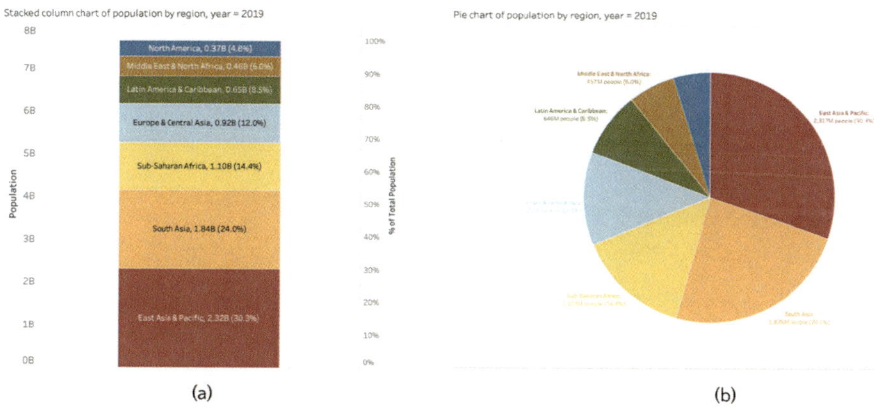

Figure 4.8. A stacked bar chart of population (a) and its corresponding pie chart (b)

We know that there are 360 degrees in a circle, so we simply divide the circle into sectors, or slices, whose central angles are proportional to their respective percent of the total amount. For example, if the East Asia and Pacific region accounted for 30.3% of the total population in 2019, then the pie slice for that region would have a central angle of 30.3% X 360° = 109.08°. If the South Asia region accounted for 24.0% of the total population of the world in 2019, then its slice would have a central angle of 24.0% X 360° = 86.4° Of course, a 90° angle represents exactly a quarter of a circle, or 25%, so the light orange slice in Figure 4.9 accounts for just less than a quarter.

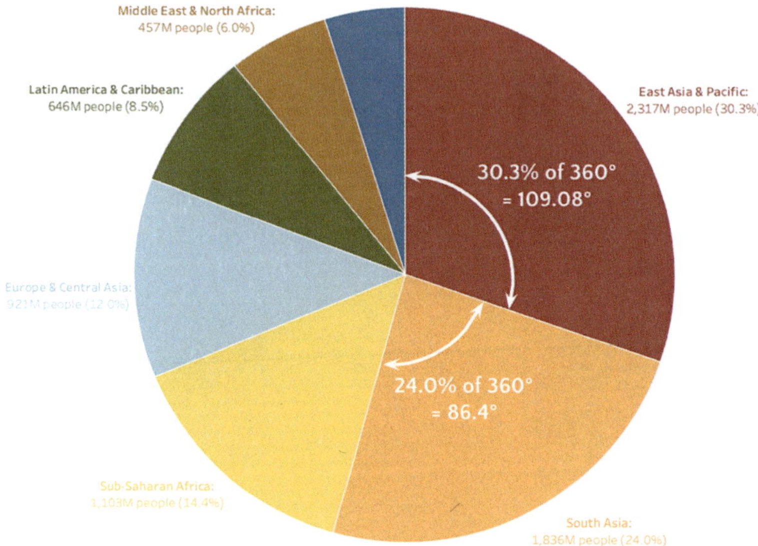

Figure 4.9. How to determine the angle of a single slice of the pie chart

The pie chart has a long history, too. The first known example of one appeared in the 1801 work *Statistical Breviary* by the very same William Playfair we've referred to once already, and it's not the last time we'll mention his name. In *Statistical Breviary*, Playfair presented data for countries in Europe at the start of the nineteenth century.

The pie charts that he included in this work represent the most significant graphical innovation contained in it. Circles with sectors had been used extensively for mathematical logic by Euler and others beginning centuries prior, but Playfair's use of one to show empirical data was revolutionary for his time.

In Playfair's work, the pie charts are actually small elements of larger and more complex infographics that contain other encodings as well. In the diagrams in the book, he used circles to represent different countries, and the area of each circle was proportional to its respective country's land area. One of the circles depicted the Turkish Empire, and he wanted to indicate that its land area was spread across three different continents: Europe, Asia, and Africa. He accomplished this by dividing the circle for the Turkish Empire into three sectors whose central angles, and thus whose areas, are proportional to the land area in each of the three continents, as shown in Figure 4.10.

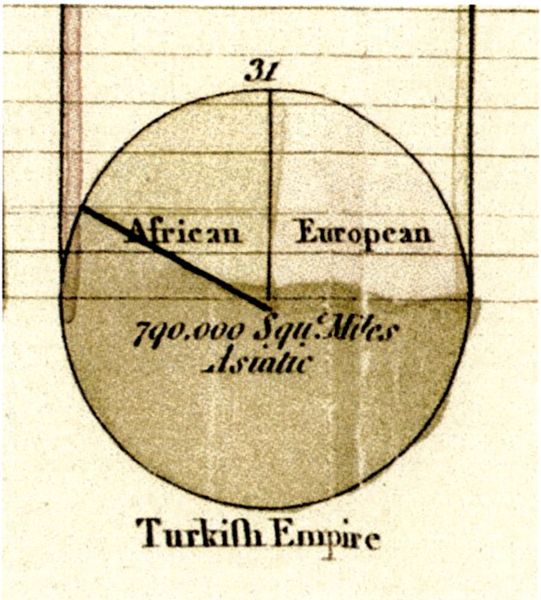

Figure 4.10. A pie chart included in Playfair's *Statistical Breviary*

In the century following Playfair's publication, it seems that pie charts became very popular, though they evidently were not called by that name. In his 1914 book, *Graphic Methods for Presenting Facts*, engineer and author Willard Cope Brinton had the following to say about what he called the "circle with sectors," comparing what we now call the pie chart shown on the right of Figure 4.11 to the stacked bar shown on the left:

> *"Fig. 2 is a form of chart used probably more widely than any other to show component parts. The circle with sectors is not a desirable form presentation, however, because it does not have nearly such flexibility as the method shown in Fig. 1. The sector method does not permit of [sic] convenient arrangement of names for the different components."*[32]
>
> –Willard Cope Brinton, *Graphic Methods for Presenting Facts* (1914)

32. https://archive.org/details/graphicmethodsfo00brinrich/page/6/mode/2up

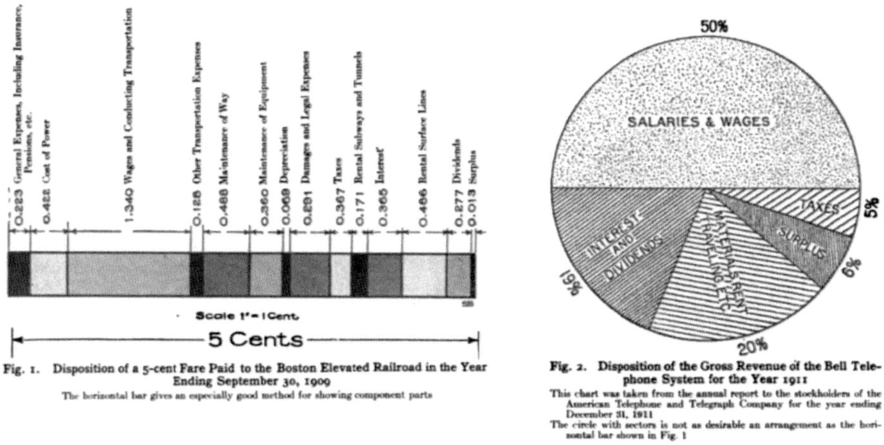

Figure 4.11. Comparing two charts in from Brinton's *Graphic Methods for Presenting Facts*

By the 1960s, the sector chart had earned its culinary metaphor nickname, but it was still receiving plenty of negative press by expert data visualization authors. In her 1969 work *Practical Charting Techniques* American graphic analyst Mary Eleanor Spear, the source of the inspirational quote for this course, had this to say about the pie chart:

> *"The pie or sector chart is a circle whose area is divided into component parts. This type of chart makes a comparison of these segments and shows their relation to the whole. It should be used sparingly, especially when there are many components. It is not only difficult to compare area segments accurately, but most difficult to label them properly."*[33]
> –Mary Eleanor Spear, *Practical Charting Techniques* (1969)

Spear agrees with Brinton's comment about the difficulty of labeling pie charts, but she adds two new cautions about the use of pie charts: first, that many components can be a problem, and second, that comparing segments accurately is difficult. Let's consider these two critiques one by one. The first point Spear makes about the use of pies when the whole has many parts can be demonstrated by the pie chart in Figure 4.12. This

33. https://www.amazon.com/Practical-Charting-Techniques-Eleanor-Spear/dp/0070600104/

figure shows the same letter usage frequency data that we considered in the section about heatmaps. This time, we consider the frequency for the English language only using a pie chart with 26 slices.

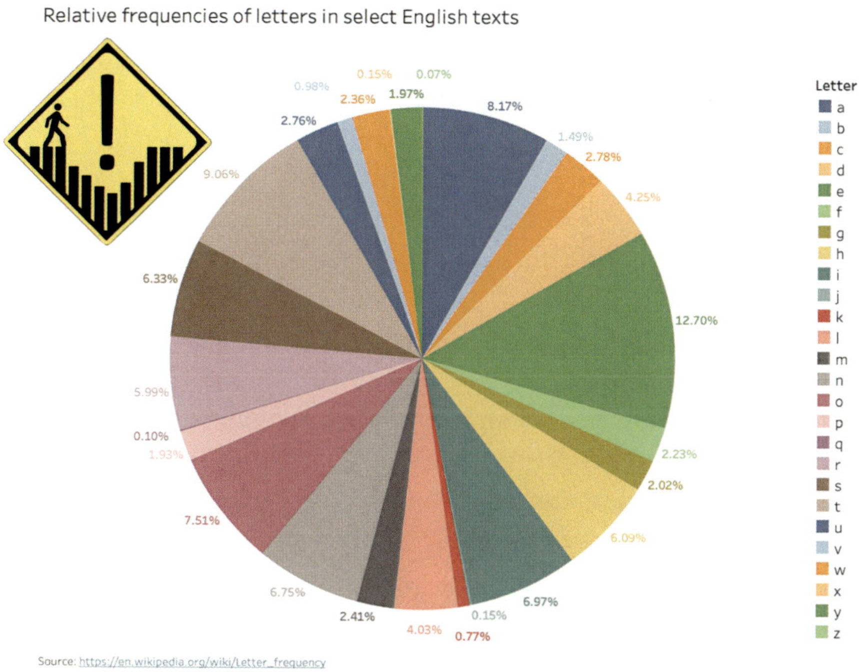

Figure 4.12. A problematic pie chart with more slices than unique color hues

The first issue with this chart is that it can feel visually overwhelming to some people. There's a lot going on with the large number of slices and color hues. A further issue is that it's difficult to compare the size of the slices at a glance. One needs to read the labels to see which of two slices that are close in size is actually larger.

A further issue is that there are 26 slices—one for each of the 26 letters of the alphabet—but only 20 distinct color hues are used. This means that some hues are repeated in the pie. Look at the legend and notice that the blue color for the letter a is the same blue hue that is used to represent the letter u. Looking at the pie chart, does the letter a correspond to 8.17% or 2.76%? The light blue for the letter b is the same as the light blue for v, so is b 1.49% or 0.98%?

In fact, the last six letters of the alphabet repeat the color hues of the first six letters, making it possible that some might misread the chart entirely, matching up the wrong letter with the wrong slice and label. This could potentially be addressed either by increasing the number of hues to 26 or by adding the letters to the slice labels themselves. We can see, however, that the chart labels are already getting cramped in places, and adding more colors doesn't sound like a recipe for improvement.

Due to these limitations, the pie chart in Figure 4.13 might be fun and colorful eye candy for someone with only a general interest in the topic, and it certainly expresses the notion of part-to-whole, but it's not a very effective visual aid for comparing part-to-part. As Spear noted, pie charts can make it difficult to compare area segments accurately. If making accurate comparisons between the different parts of the whole is important, it might be advisable to choose effectiveness over expressiveness and use the bar chart shown in Figure 4.13. Note that the percent of usage of each letter encodes both the heights of the bars as well as their color saturation. This is another redundant encoding because we can use both of them to make the exact same comparison.

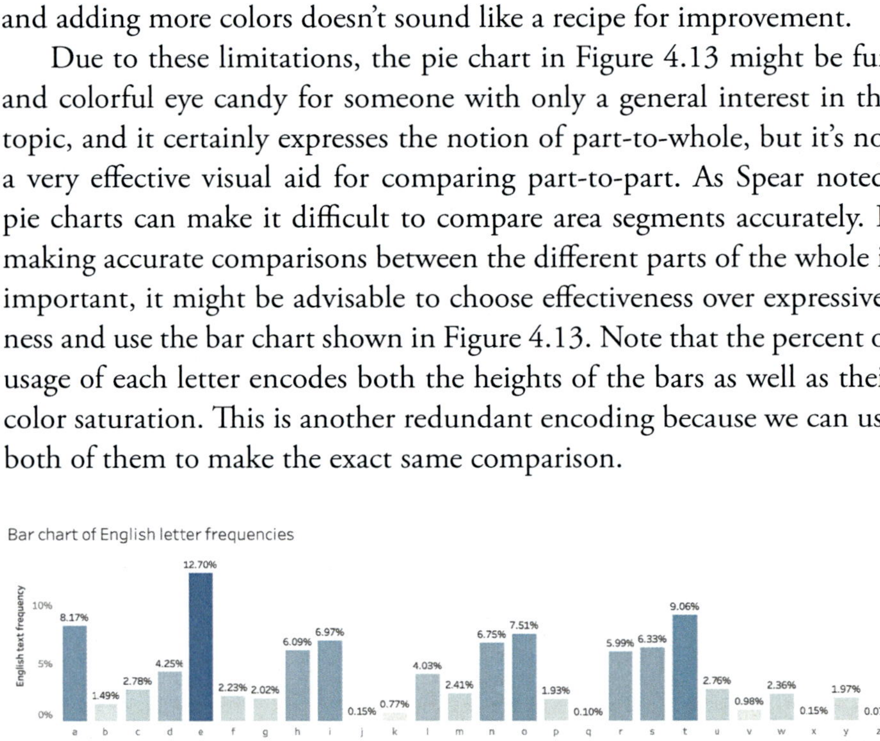

Figure 4.13. A bar chart showing letter usage frequency for English texts

The diagram shown in Figure 4.14, created by Wikipedia user Schutz, does an excellent job showing that comparison of pie slices is more difficult for us than comparison of bar lengths. In this diagram, three different sets of percentages are used to create pairs of charts: a pie chart and its corresponding bar chart below it. When we look at the pie charts, it's next to impossible to determine the difference between them, and what relationship might exist between the slices of each pie. When we look below at the bar charts, we can tell immediately that the first data set has values that get successively larger, the middle one has values that are roughly the same, and the third one has values that get successively smaller.

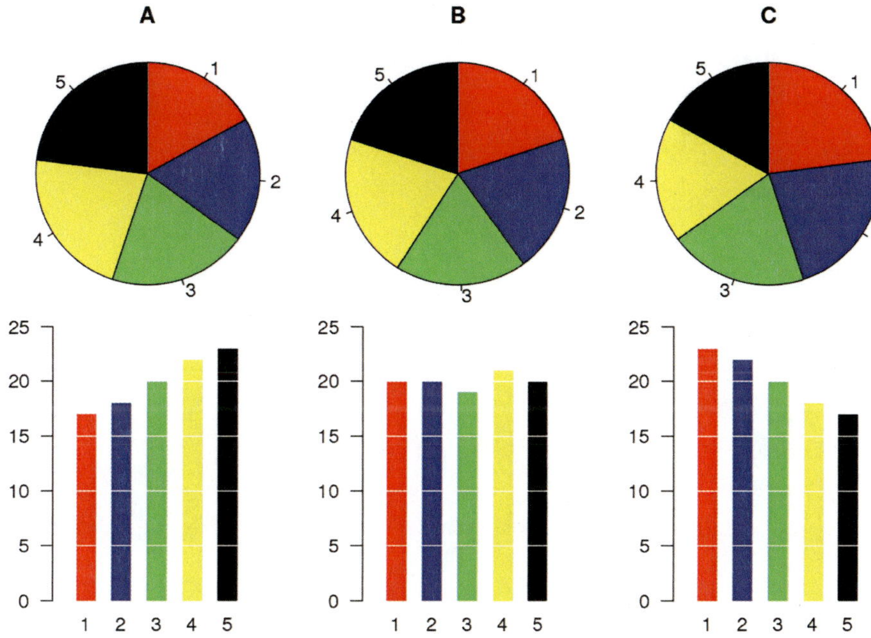

Figure 4.14. Three sets of percentages, plotted as both pie charts and bar charts[34]

Once again, the bar chart, a chart type that *doesn't* naturally express a part-to-whole relationship, is superior to one that *does* express part-to-whole, the pie chart. So it would seem that we should listen to and agree with the data visualization experts who don't like pie charts, including those in more recent years who have proclaimed "pie charts are evil," or who have even called for "death to pie charts." Banish pie charts for good, right?

Not so fast. Circles, according to Manuel Lima, founder of VisualComplexity.com, are "the universal symbol of unity, perfection, movement, and infinity."[35] We talk about the "circle of life" in nature, the "circle of fifths" in music, and a "circle of trust" in social relationships. We should definitely be aware of their shortcomings as visual encodings

34. CC BY 1.0, https://commons.wikimedia.org/wiki/User:Schutz#/media/File:Piecharts.svg
35. https://www.amazon.com/Book-Circles-Visualizing-Spheres-Knowledge/dp/1616895284/

of data, but we should not be so quick to abandon them. All charts have their drawbacks, and the pie chart is no different.

We can think of instances in which the pie chart wouldn't just suffice, but it might actually be a better way of visualizing the data than a bar chart. Let's say, for example, that we wanted to make the point to an audience that the population of South Asia was just less than a quarter of the population of the entire world. Take a look at Figure 4.15 and compare the bar chart, the stacked bar chart, and the pie chart.

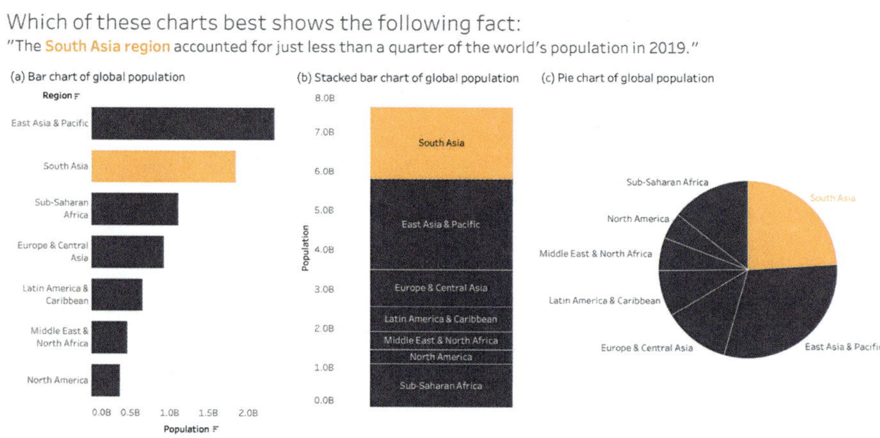

Figure 4.15. The pie chart can be effective at showing percent of total

One could make the argument that the pie chart is the most effective out of these three options at making this specific point. Since the slice for South Asia starts at the 12 o'clock position and ends very close to the 3 o'clock position of a standard clock face, it's easy to see that it takes up about a quarter of the total area of the circle. This fact isn't easy to see when looking at the other two charts.

So we see that there are cases when a pie chart wouldn't be very helpful at all, and there are cases when it would be quite helpful. The same can be said of every type of chart. Much of the frustration comes from the fact that pie charts are often used in unhelpful ways, it can't be denied. But there's no reason to throw the baby out with the bath water. The best quote about pie charts comes from American statistician and computer scientist Leland Wilkinson:

"A pie chart is perhaps the most ubiquitous of modern graphics. It has been reviled by statisticians (unjustifiably) and adored by managers (unjustifiably)." —Leland Wilkinson, *The Grammar of Graphics* (p. 23)[36]

There are two other common problems with pie charts of which to be aware. Since the chart forms a complete circle, the sectors should be mutually exclusive and collectively exhaustive. This forms the acronym MECE, which is pronounced "ME-see."[37] The first part of the acronym, "mutually exclusive," means that an individual item or data point should belong to one and only one of the slices. If the slices of the pie aren't mutually exclusive, then it's possible that the sum of the percentages of the slices adds up to more than 100%.

For an illustrative example of this problem, take a look at the pie chart in Figure 4.16, which shows the results of a YouGov poll asking voters which of a list of candidates they would be enthusiastic about supporting in an upcoming election.

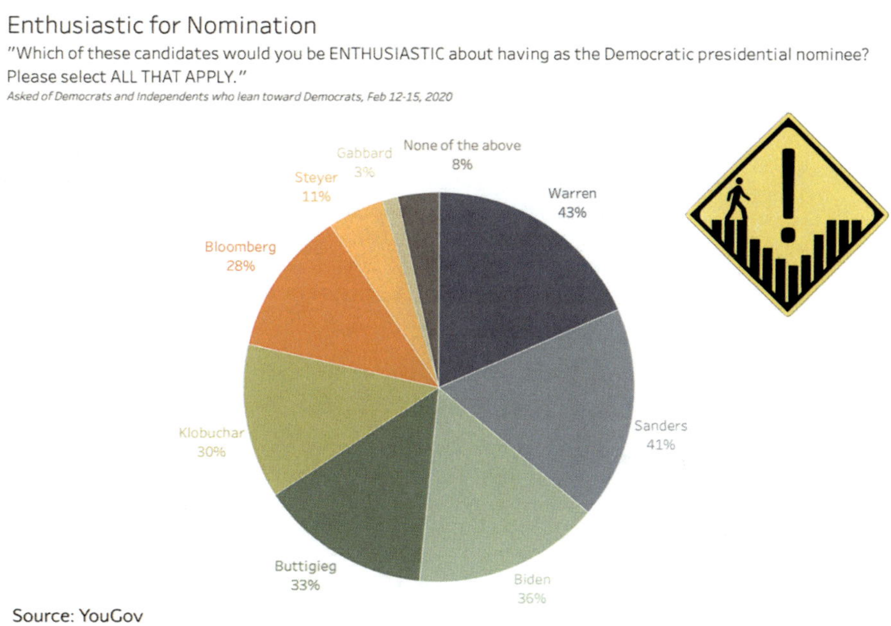

Figure 4.16. A pie chart where the slices add up to more than 100%

36. https://www.amazon.com/Grammar-Graphics-Statistics-Computing/dp/0387245448/
37. https://en.wikipedia.org/wiki/MECE_principle

At first blush this pie chart might seem innocent enough, but a closer look reveals that the slices aren't nearly as large as their labels suggest they should be. We see that 43% of respondents at that time said they'd be enthusiastic about supporting Elizabeth Warren, but her slice in the pie comprises less than 20% of the circle. The other slices have the same issue. If we add up the labels of the slices, we get 233%. What's going on here?

The cause of this issue is that respondents were able to select more than one candidate in the poll. The choices weren't mutually exclusive. This occurs often in surveys where checkboxes allow survey takers to "select all that apply." It's very misleading when this type of data is shown as a pie chart. Instead, the bar chart would be a better representation for this data, because it makes no false promises about part-to-whole relationships. See Figure 4.17 for a bar chart version of this exact same data.

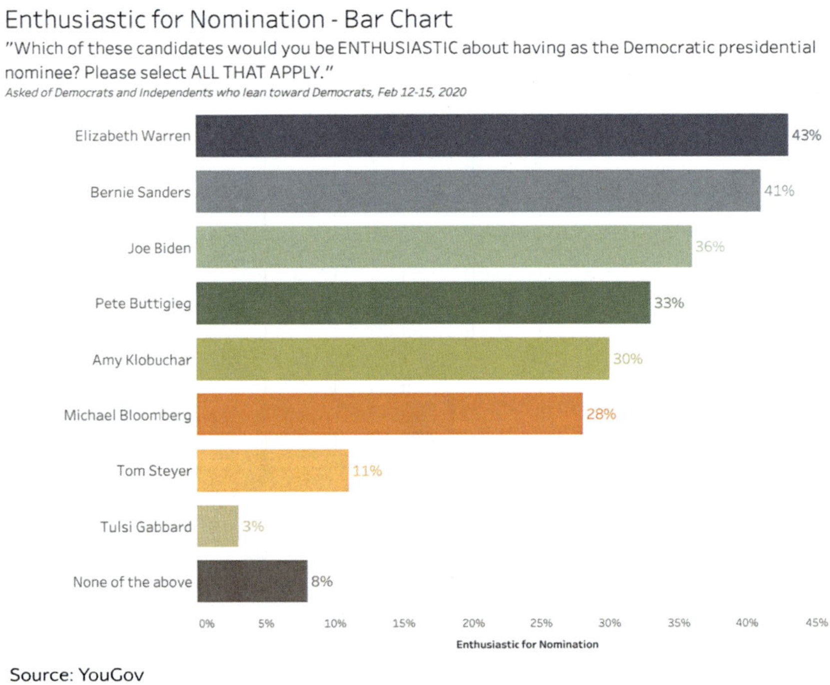

Figure 4.17. A bar chart of voter enthusiasm by candidate

The second part of the MECE principle stands for "collectively exhaustive." If we were to create a pie chart out of the populations of an *incomplete* selection of countries, regions, or income groups, then this could be misleading if the populations don't form a meaningful unit or whole. Consider the two pie charts in Figure 4.18. The one on the left shows all four of the different income groups in a chart that expresses completeness. The slices are therefore collectively exhaustive. On the right side, however, this second pie chart only includes two of the four levels of the category. The slices are not collectively exhaustive in this version, so it's misleading that they form a complete circle. This practice should be avoided.

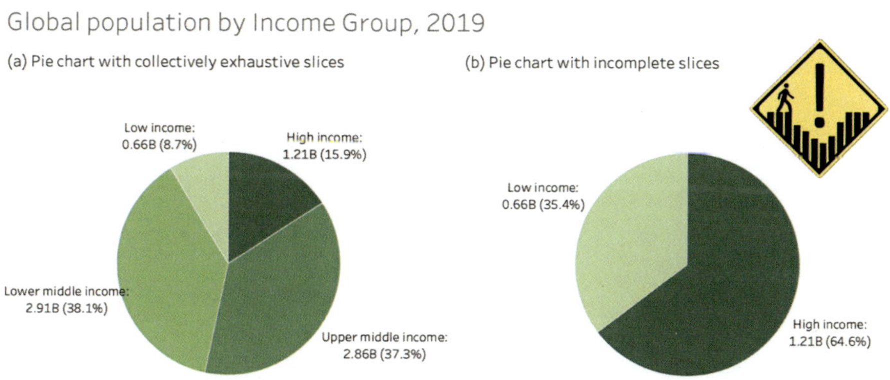

Figure 4.18. A pie chart (a) with collectively exhaustive slices and another (b) without

This leads us to our next "Chart Reading Tip":

> **TIP #5.** If the chart expresses completeness or fullness, make sure the segments are MECE: mutually exclusive and collectively exhaustive.

Before we conclude the section on the much maligned and highly controversial pie chart, let's consider one stylistic modification: the **donut chart** (or doughnut chart, depending on where you live). The donut chart is simply a pie chart with a circle removed from the center, as shown in Figure 4.19. This hole in the middle may hide the central angles of the slices, but we can still make rough comparisons using the position of the edges of the slices, their areas, and the length of the arcs they form.

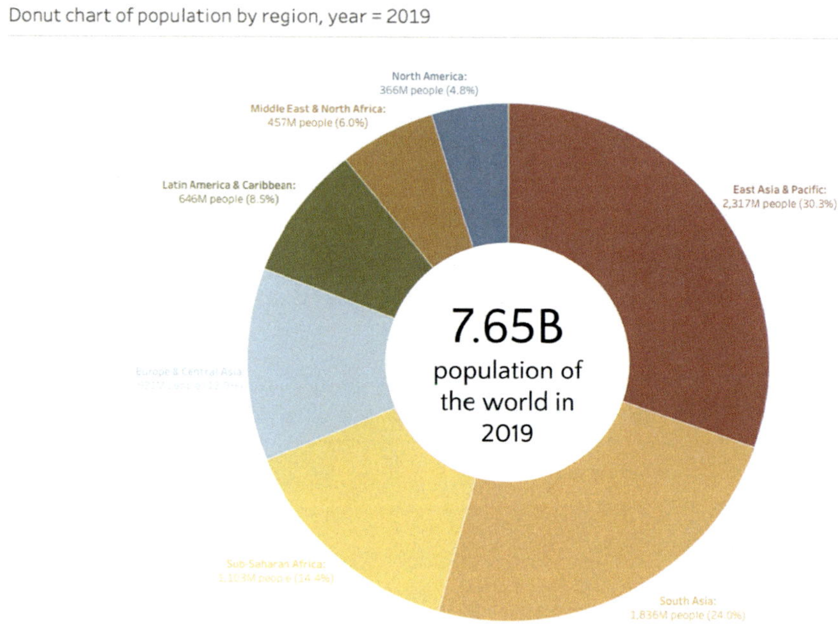

Figure 4.19. A donut chart of world population by region

As you can see in Figure 4.19, the hole in the center leaves a convenient space for other information such as individual data points, logos, or other graphics that are related to the data in the donut chart.

The Treemap

So far we have broken down the world's population into regional values by dividing a bar into segments (the stacked bar chart) and by slicing a circle into sectors (the pie chart). These charts expressed the part-to-whole relationship using, respectively, length and angle encodings. Technically, though, we also use the area encoding in both of these charts to compare relative populations of the regions. A slice that has twice the area of another in the pie has twice the population, and a rectangle that has twice the area of another rectangle in the stack similarly has twice the population.

We know from basic geometry that the area of a rectangle is equal to its length times its width. Since the widths of the segments in a stacked bar chart are all equal, the differences in their area is due to differences in length only. But what if we allowed the rectangles to have different lengths *and* different widths, and encoded the population using only area? And what if the resulting segments all fit together to form a larger rectangle, thereby expressing unity or fullness? The resulting chart is called a treemap, and we can see what it looks like in Figure 4.20.

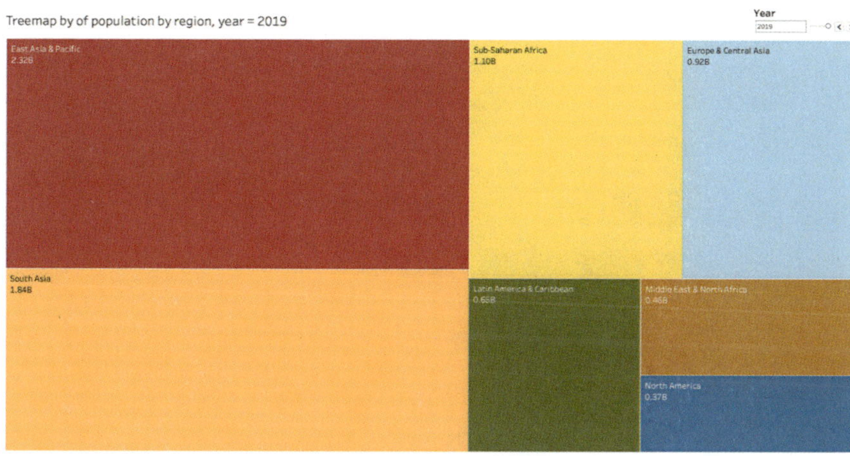

Figure 4.20. A treemap of world population by region

You may be wondering why a chart that's shaped like a quilt has the word "tree" in its name. This chart uses the area of each rectangle to encode the population of each region. Since the 2D area channel ranks in the middle of Munzner's effectiveness list, it seems at first blush like this chart type doesn't give us any advantages over the others we've already considered in this chapter. That's because this version of the treemap is actually too simple to put its virtues on display. To really shine, a treemap needs hierarchical data.

A hierarchy is defined as "any system of persons or things ranked one above another." A hierarchy is often visualized using a tree structure with branches and leaves, such as a diagram showing the folders and files on a computer's hard drive. This is exactly the situation that the chart's inventor, Ben Shneiderman, was trying to visualize when he was hit with the idea to create one while working at the Human-Computer Interaction Lab of the University of Maryland in 1990.[38]

38. http://www.cs.umd.edu/hcil/treemap-history/

Our population data also contains multiple levels of a hierarchy in it: the world is broken down into regions, which are in turn broken down into countries. Figure 4.21 illustrates this property of the data in tree form.

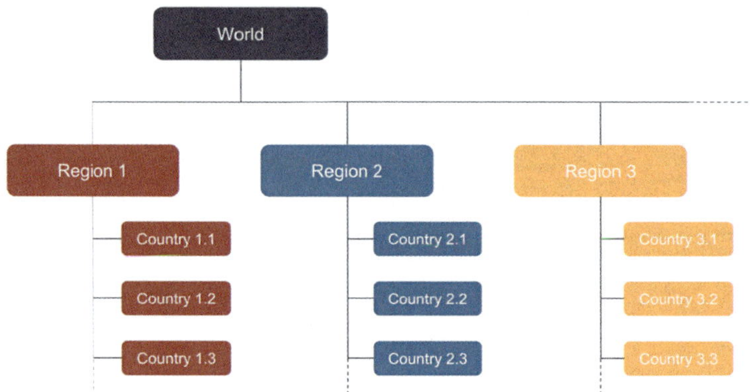

Figure 4.21. A hierarchical structure within our population data set

With this hierarchical property in mind, we can add more detail to the treemap by breaking down each region's rectangles into multiple rectangles, one for each country in the data set. In this way, we can fit all of the 217 countries into the view with no need to scroll, even if some of them are so small that we can't really see them. Figure 4.22 shows the treemap of region and country population.

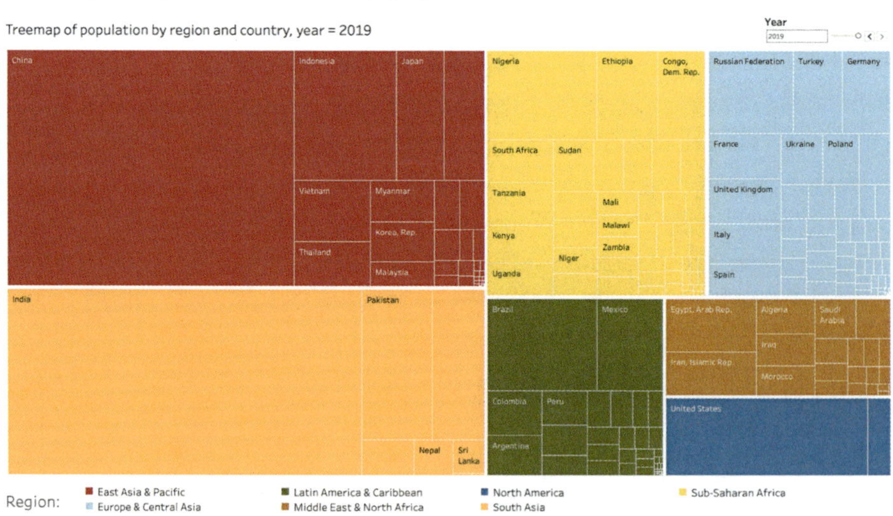

Figure 4.22. The treemap when we add the country level of the hierarchy

Notice how the white boundaries between the regional clusters are thicker than the white boundaries between the country rectangles within each regional subgroup. This results in what's often called "nesting."

Since many of the rectangles in the treemap are too small to include their labels, this type of chart is often best viewed on a screen with "tool-tip" capabilities that provide a pop-up callout when the reader hovers a mouse cursor or taps a touchscreen.

The way each region is colored with its own hue, as shown in the legend at the bottom of the chart, makes the thicker regional divisions very easy to see. But there's another way that color can be used in a treemap. Instead of using color hue to accentuate the different levels of the hierarchy, we can use color saturation to redundantly encode the quantitative variable used to size the rectangles—in this case, population. Figure 4.23 shows this version of the treemap, with larger rectangles such as China's and India's having more saturated blue shadings. The legend at the bottom helps us understand this redundant usage of color.

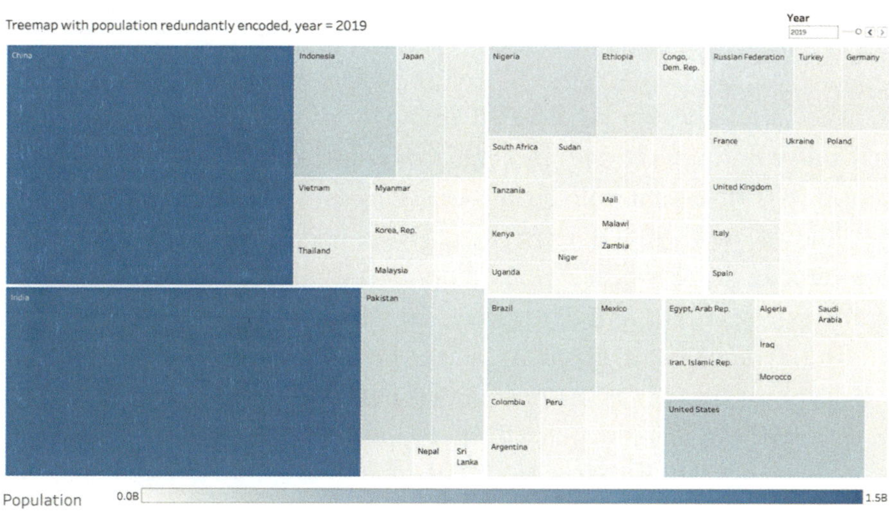

Figure 4.23. A treemap with color saturation redundantly encoding population

A very powerful application of the treemap, however, involves the use of a different quantitative variable to encode color, instead of using color redundantly. For example, if we wanted to compare population size with life expectancy, we could size the rectangles by population again, and this time we could color the rectangles using the life expectancy variable. Figure 4.24 shows this version of the treemap, which uses data from 2017 since that was the most recent year that life expectancy figures were available on the World Bank data site.

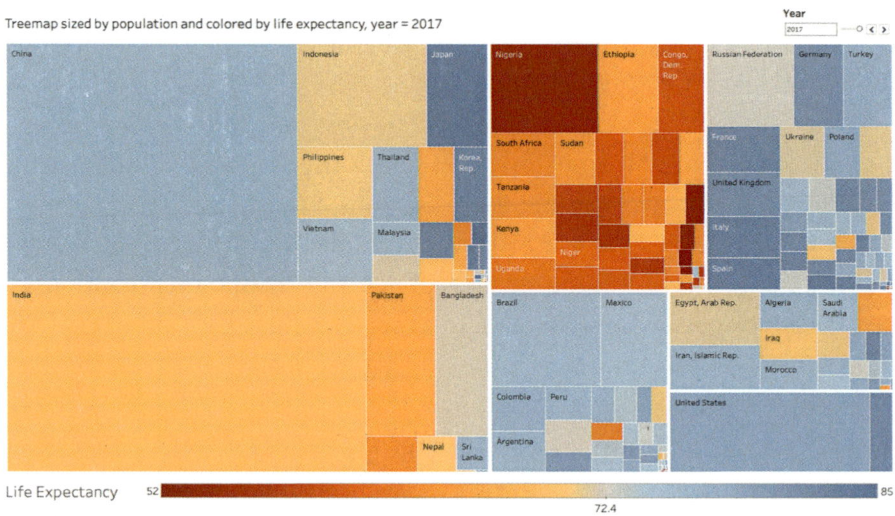

Figure 4.24. A treemap with different quantitative variables for size and color

Take a look at the legend at the bottom of this treemap and notice that this colormap changes in color from orange to blue. This is what's known as a divergent colormap, because color gradually diverges from one hue to another over the range. The countries with the lowest life expectancies are the most saturated orange and the colors with the highest life expectancies are the most saturated blue.

This type of color encoding works best when there's a meaningful middle point that can be used as a pivot from one color hue to the other. In this case, the color switches from orange to blue at a life expectancy of 72.4, which, according to the World Bank, was the overall global life expectancy in 2017. Since the color encoding diverges at this exact point, it's easy to tell by looking at the treemap which rectangles have life expectancies below the global level, and which have life expectancies higher than the global level.

When using a diverging color palette, care needs to be taken to avoid printing or otherwise looking at the chart in black and white, as the darker orange shapes will look identical to the darker blue ones, and the encoding will mislead the reader. Care also needs to be taken in many cases to avoid a color encoding that diverges from red to green, as this form of color vision deficiency affects up to 8% of males and 0.5% of females of Northern European descent.[39] Of course, if this is not your audience, or if the audience is small enough that you can check for color blindness ahead of time, then feel free to ignore this warning.

39. https://en.wikipedia.org/wiki/Color_blindness

FURTHER LEARNING

- Pie charts:

The Book of Circles: Visualizing Spheres of Knowledge by Manuel Lima: https://www.amazon.com/Book-Circles-Visualizing-Spheres-Knowledge/dp/1616895284/

"An Illustrated Tour of the Pie Chart Study Results" by Robert Kosara: https://eagereyes.org/blog/2016/an-illustrated-tour-of-the-pie-chart-study-results

- Treemaps:

"Treemaps for space-constrained visualization of hierarchies" by Ben Schneiderman, Later updates by Ben Shneiderman and Catherine Plaisant: http://www.cs.umd.edu/hcil/treemap-history/

Data Stories podcast episode 029: "Treemaps w/ Ben Shneiderman" (2013) https://datastori.es/data-stories-29-treemaps-w-ben-shneiderman/

"Tree-Maps: A Space-Filling Approach to the Visualization of Hierarchical Information Structures" by Brian Johnson and Ben Shneiderman (1991) http://www.cs.umd.edu/hcil/trs/91-06/91-06.pdf

Chapter 5

SEEING CHANGE OVER TIME

"We think about time as movement on a line in space."
—Barbara Tversky

Up until this point, all of the charts we've considered have displayed data that was pulled from one period of time—and only one. We've looked at charts with dots or bars or pie slices that encode country population figures from the year 2019, or life expectancy estimates from 2017. Look at these charts again and ask yourself if you can figure out how they relate to time:

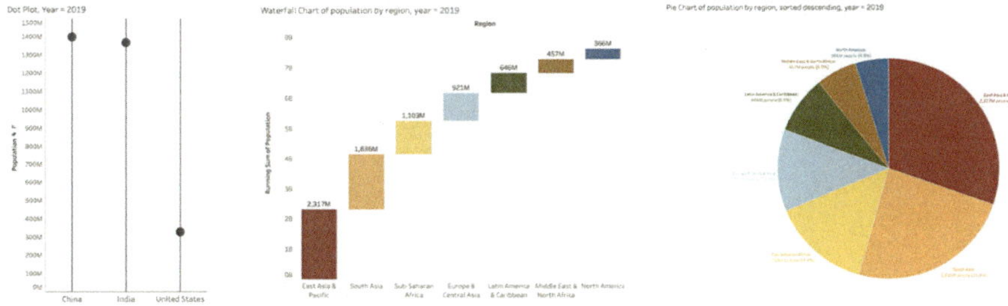

Figure 5.1. Three charts that use data from one and only one point in time

These graphical displays all show data attributed to the same time (a single year), but none of them give us any sense of how these quantities have *changed* over time. Has the variable been on the rise, on the decline, or fairly steady in recent years? We don't know because the charts only show data for one point in time, like a photograph that captures a fleeting moment.

Mapping Time to Space

Only not quite, because even a photograph can give a clear sense of movement and change over time. For example, consider this photograph of Matt playing frisbee with his dog, Henry VIII. See how your mind fills in the past and anticipates the future based on this one instant in time:

Figure 5.2. Matt playing frisbee with his dog, Henry. Photo credit: Samantha Jones

What happened the split second before the photograph was taken, and what happened immediately afterward? Unless you have aphantasia, a mental condition characterized by an inability to voluntarily visualize mental imagery, you likely imagined something similar to the images on the left and right of the original image in the middle in the following series:

Figure 5.3. A photograph can imply a past and suggest a future

We can't really do this with a bar chart showing population for one year, or sales by region for one quarter or one month. We don't have any idea where these quantities have been or where they're going based on the kinds of charts we've considered so far.

But any understanding of where things are today needs to be accompanied with an understanding of where they have been in the past in order to give us a full picture. Nothing is static in our world. As the ancient Greek philosopher Heraclitus of Ephesus famously stated, "Everything changes and nothing stands still." Or, as he said elsewhere, "You cannot step twice into the same river."

The moment you step into the river, the river changes, and it will never be exactly the same. Data is like a river. As soon as we gather data, the universe from which it was taken shifts, and then it immediately shifts again.

Having our finger on the pulse where things *are*, where they *have been*, and where they *might be going* is core to our ability to navigate our time here on the planet. This is why graphical displays of data often encode the all-important variable of time. We need a way to see the change that has taken place, so we often map time to space, or position, in diagrams and charts. Did you notice how often the word "where" appeared in the preceding paragraphs to describe a point in time?

How do you envision the changing of time, in your mind? Likely you think of time differently at different levels: hours in a day, days in a week, months in a year. For those of us who learned how to tell time using the standard 12-hour clock face, we know that it's 6 o'clock when the hour hand points straight down and the minute hand points straight up. And we know that at 9 o'clock, the hour hand will be in a different place, pointing directly to the left, 270° from the 12 o'clock position. In this way, the clock face maps time to space, encoding the hour, minute, and second using position on the very same polar coordinate system we discussed when we considered pie charts.

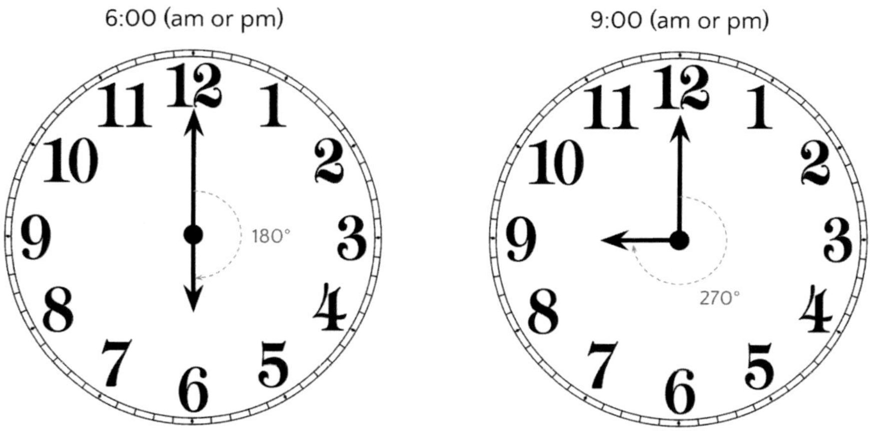

Figure 5.4. The standard clock face maps time to polar position in space

How about the way a year is broken down into seasons and months? What month is it right now, and if you imagine yourself being in this present month, what does that look like in your mind? Do any graphical forms come to the surface for you?

If you think of some future moment—say, six months from now—*where* do you picture that future in your mind, relative to where you are now? Is it in front of you or above you or somewhere else? Do you see the pages of a calendar flying by? Do you picture yourself traveling on a straight line, or going around a racetrack that loops back on itself when December turns into January, or perhaps spiraling upward or downward? These are all mappings of time to space, and we do not all map time to space the same way.

Here's another mental exercise for you to explore your own perception of time. Imagine a square piece of paper, and imagine three dot-shaped stickers. The first sticker is placed in the very center of the sheet of paper, and you're told that this first sticker represents "lunch." You're then asked to place the two remaining stickers anywhere you want on the page. One of them represents "breakfast" and the other represents "dinner." Where do you put them?

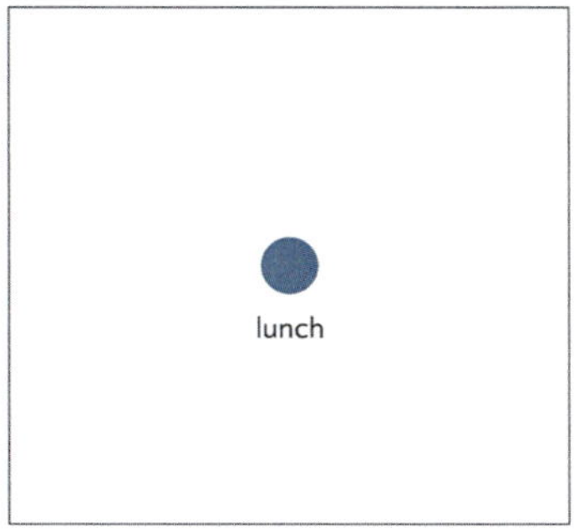

Figure 5.5. An exercise of mapping time to space: place dots for breakfast and dinner

Researchers Tversky, Kugelmass, and Winter asked this very question of children and adults of different backgrounds in experiments they wrote about in 1991.[40] They were interested in understanding how the participants would map time to space on the page. Many put the stickers in the order that the meals occur in a day from left to right, but not all of them did.

Interestingly, according to the researchers, "left-to-right was dominant for speakers of English and right-to-left for speakers of Arabic, with Hebrew speakers in between." Of course, English is a language that's written from left to right, Arabic text goes the opposite way, from right to left. Interestingly, Hebrew words are written right to left, while numbers are written left to right.

English	**Time flies over us...** →
Arabic	← ... الوقت يمر فوقنا
Hebrew	← ... הזמן עף עלינו

Figure 5.6. The beginning of a Nathaniel Hawthorne quote in three different languages

40. Tversky, B., et al. "Cross-cultural and developmental trends in graphic productions," *Cognitive Psychology* 23 (1991): 515–557.

So if you grew up with English as your first language, it's likely that you imagined putting the breakfast sticker on the left of the lunch sticker, and the dinner sticker on the right, in a straight, horizontal line. Did you notice that the sequence of photographs of Matt and Henry in Figure 5.3 is in this exact left-to-right orientation? The "before" photograph is to the left and the "after" photograph is to the right.

But the direction of *motion* in each individual photograph points from right to left—the direction that the frisbee is flying and Henry is jumping. So it's likely that reversing each photograph while keeping the sequence order the same, as in Figure 5.7, feels more "natural" to those of us with English-speaking backgrounds. In this version, both motion and sequence proceed in the same direction as these very words you're reading right now:

Figure 5.7. Mirrored versions of each photograph in the same left-to-right sequence

What does all of this mean for how we see changes in data over time? It means that if your original language is a left-to-right language such as Italian, or Spanish, or English, then there's a good chance that the most natural expression of time for you is a horizontal line, with time progressing from left to right. Charts that use this convention for time will satisfy the expressiveness principle for you, at least with respect to time. It's important to note that other chart readers of different cultures might not see it that way.

In the rest of this chapter, we'll focus on this left-to-right temporal orientation in the charts we review, knowing that this convention is not a universal one.

The Line Chart

So left to right is correlated with increasing time for many chart readers, but is this equally true for all mark types, or are some mark types more expressive of change over time than others? Researchers from Stanford studied the way people interpret the meaning of bars compared to lines, and they found "a strong tendency to describe data portrayed as bars discretely (e.g., 'A is higher than B') and to describe data portrayed as lines in terms of trends (e.g., 'X increases from A to B')."[41]

What does this difference between bars and lines mean? Take a look again at the bar chart of country population we considered earlier, and compare it to the version that instead connects the population dots with straight lines in Figure 5.8. Notice that when you look at the version with the lines, assuming you have a left-to-right temporal perspective, your mind wants to tell you that something is going *down* or *decreasing* along with the downward sloping line. The bar chart, however, simply presents you with three different boxes to compare with each other. These boxes don't express continuity in quite the same way. This isn't to say that we can't ever use bars to show change over time. But lines are definitely more expressive of change over time.

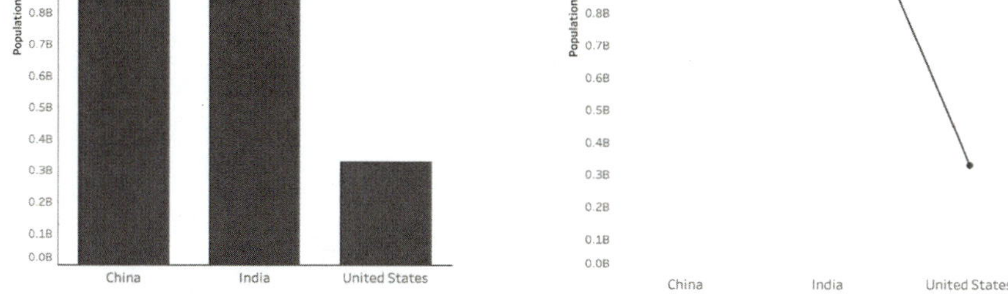

(a) (b)

Figure 5.8. Bars (left) express discrete comparisons, while lines (right) express trends

41. Zacks, J., and Tversky, B., Bars and lines: A study of graphic communication, *Mem Cogn* 27, 1073–1079 (1999). https://doi.org/10.3758/BF03201236

Furthermore, the researchers found that while time tended to fall on the horizontal axis, value tended to favor the vertical axis; specifically, higher vertical positions tend to be associated with higher value or quality. Tversky et al. found that "in general, more, better, and good are associated with up, high, and top, and less, worse, and bad are associated with low, down, and bottom."[42]

And while a person's mapping of time to position seems to be related to their written language, "higher is better" is an assumption that seems to be fairly universal, and not dependent on language or culture. We can reason that this makes sense, because each one of us would agree that working against gravity to climb a mountain is a sign of strength and vitality. Up is good.

Figure 5.9 presents a typical line chart: it expresses time progressing from left to right (horizontal position, x-axis) and value increasing from bottom to top (vertical position, y-axis).

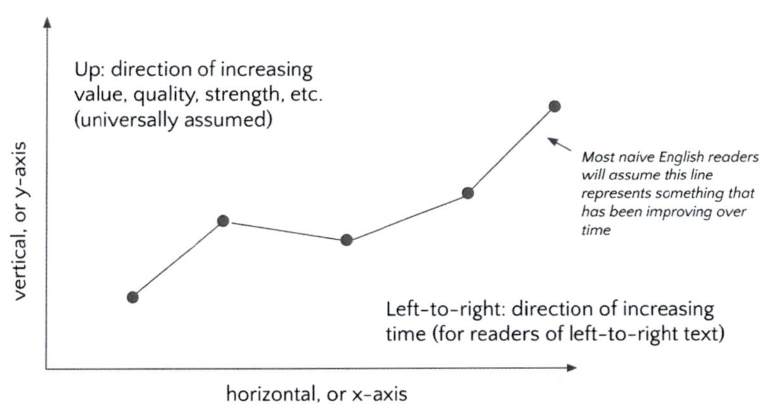

Figure 5.9. A line chart that shows common conventions for English speakers

This leads us to our next two Chart Reading Tips:

TIP #6. If a chart expresses change over time, check what direction corresponds with increasing time. It's likely left-to-right, but might not be.

42. Tversky, B. et al., "Cross-cultural and developmental trends in graphic productions," *Cognitive Psychology* 23 (1991): 515–557.

TIP #7. If a chart expresses value or worth, check which direction corresponds with "more," "better," or "improvement." It's likely up, but might not be.

These tips take us all the way back to 1786 and the innovations of William Playfair once again. It turns out that Playfair's invention of the line chart followed these exact conventions. Remember the bar chart of imports and exports to Scotland that we considered back in Chapter 3, shown in Figure 3.7? What's interesting about that bar chart is that it was the only bar chart in the series of 44 charts in the publication *The Commercial and Political Atlas*. The other 43 charts were all line charts, an example of which is shown in Figure 5.10.

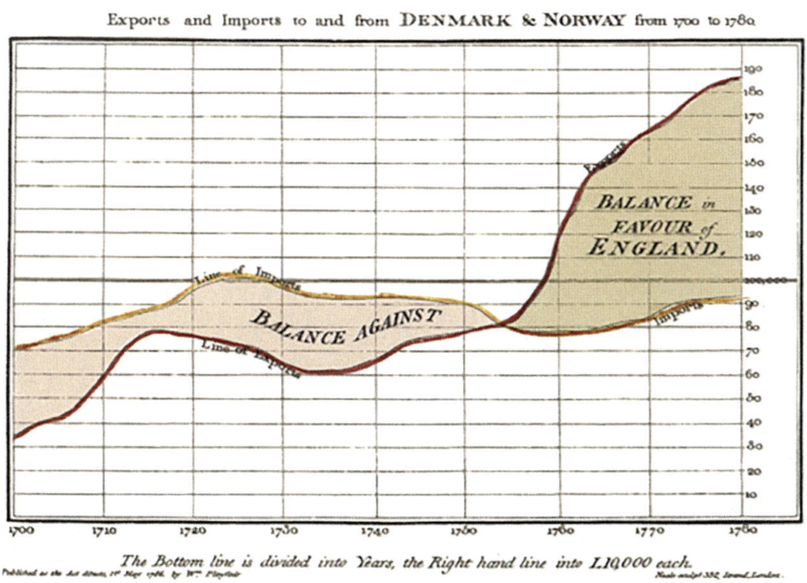

Figure 5.10. William Playfair's line chart (Public domain, via Wikimedia Commons)

So why did Playfair include 43 line charts and only one bar chart in his publication? The answer is pretty simple: he only had data for Scotland for one single year. For all of the other countries, he had data for multiple years, so he was able to show how imports and exports had changed over time. For Scotland, his own country of birth, he had to fall back on a "snapshot" view, and so the bar chart was born. As the popular saying goes, "Necessity is the mother of invention."

Playfair summarized his own assessment of the relative value of the bar chart as compared to the line chart in the following quote:

> *"This Chart [the bar chart]...does not comprehend any portion of time, and it is much inferior in utility to those that do."*
> — William Playfair, *The Commercial and Political Atlas,* 1786

If charts that show change over time are so valuable, then let's build our own simple chart that shows change over time, and let's stop to consider the lines carefully. If we start with a dot whose vertical position encodes the population of the world in 2019—7.646 billion, according to the World Bank—and then if we similarly add dots for each year going back to 1960, we get the following dot plot:

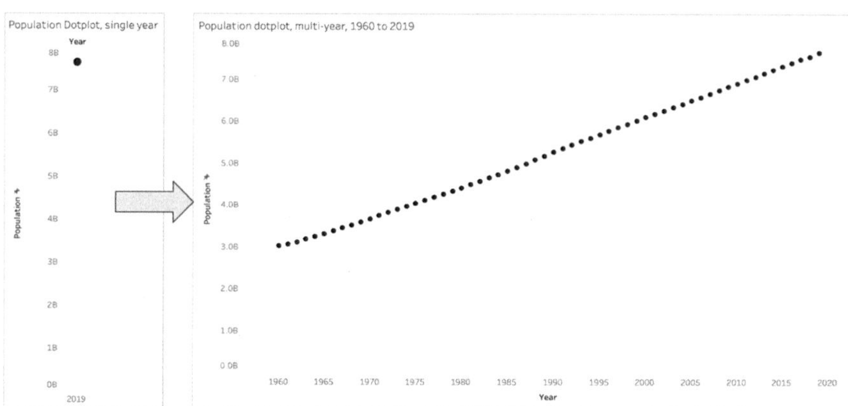

Figure 5.11. Two dot plots: one for one year and one for a half century

How do we connect these dots? Do we even need to connect them? The motion seems implied. We can tell that the population of the world has been increasing in fairly consistent annual amounts for the past six decades. We can leave the dot plot as is if we want, or we can take advantage of the fact that our brain will interpret connecting lines as implying a relationship or trend from one point to the next. Figure 5.12, offers four different ways to add lines to the dot plot.

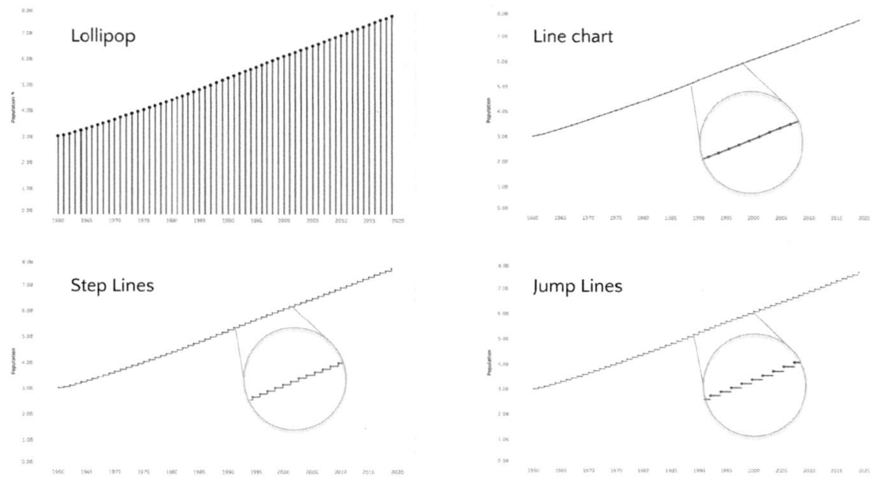

Figure 5.12. Connecting the dots with different types of lines

The first way of adding lines yields the lollipop chart that we have already considered. This technique does not actually connect the dots to each other at all, but rather connects each of the dots to the baseline. The second way of adding lines is perhaps the simplest: just connect a dot to its adjacent dots with a perfectly straight line. This yields our basic line chart.

The next two methods, step lines and jump lines, might seem a little strange. They're certainly less common than the basic line chart. In the case of the step line chart, a horizontal line extends to the right of any particular dot (except the last one in the series), and when it gets to the horizontal x-position of the next dot, it turns at a right angle either up or down and extends to the next dot in the series. Jump lines are similar but they lack the vertical line as the horizontal line comes to an end at the x-position of the adjacent dot and then "jumps" to the next dot and extends to the right in similar fashion.

Which of these four methods is more appropriate for showing a changing population? While the data are collected once per year, we can reason that the population of the world continually changes day by day, moment by moment, as some die and others are born. The actual path between two adjacent annual readings might not be perfectly linear, but it's probably fairly close to linear. So connecting the dots to each other with a straight line as shown in the line chart makes good sense.

When would step lines or jump lines be a better choice for connecting the dots? Scenarios when the value being charted stays constant for a period of time and then changes all of a sudden, in one single moment, are perfect for step and jump lines. For example, consider the step line chart showing the change in stamp prices over time in the United States shown on the left-hand side of Figure 5.13.

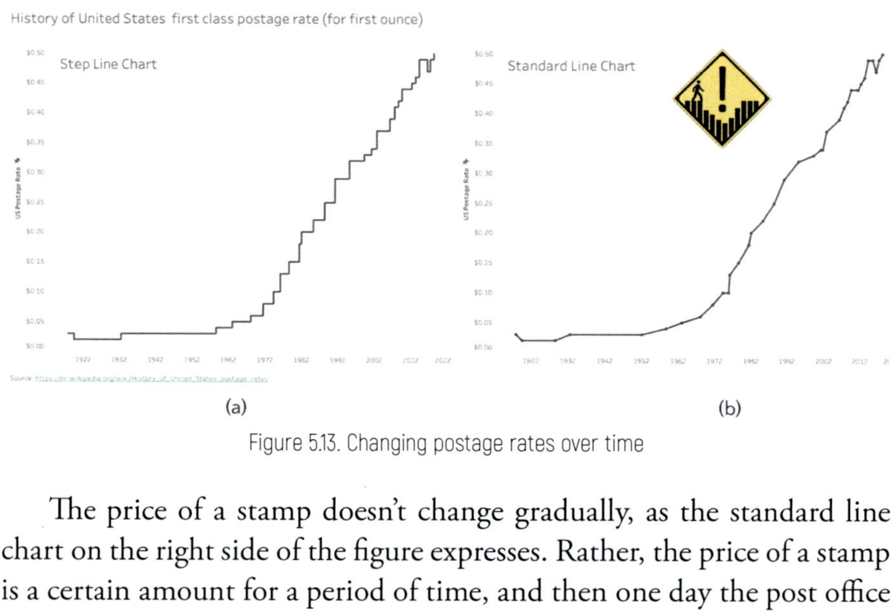

(a) (b)

Figure 5.13. Changing postage rates over time

The price of a stamp doesn't change gradually, as the standard line chart on the right side of the figure expresses. Rather, the price of a stamp is a certain amount for a period of time, and then one day the post office opens for business and the price has changed to a new value that will itself remain in effect for a while. It will stay that same new price for many days, and then change again in stepwise fashion on some later day. So connecting the dots with a straight line would be confusing, misleading, and technically inaccurate. We could consider it a violation of the expressiveness principle because the price of a stamp doesn't actually ramp from one value to the next.

Similarly, world records are set by an athlete or performer in one moment, held for a period of time, and then suddenly replaced with a new world record when it's broken. The men's mile run world record shown in Figure 5.14 is therefore best shown using jump lines. A standard line chart would be misleading, because the record didn't gradually change over time from one record to the next; it changed suddenly, and all at once.

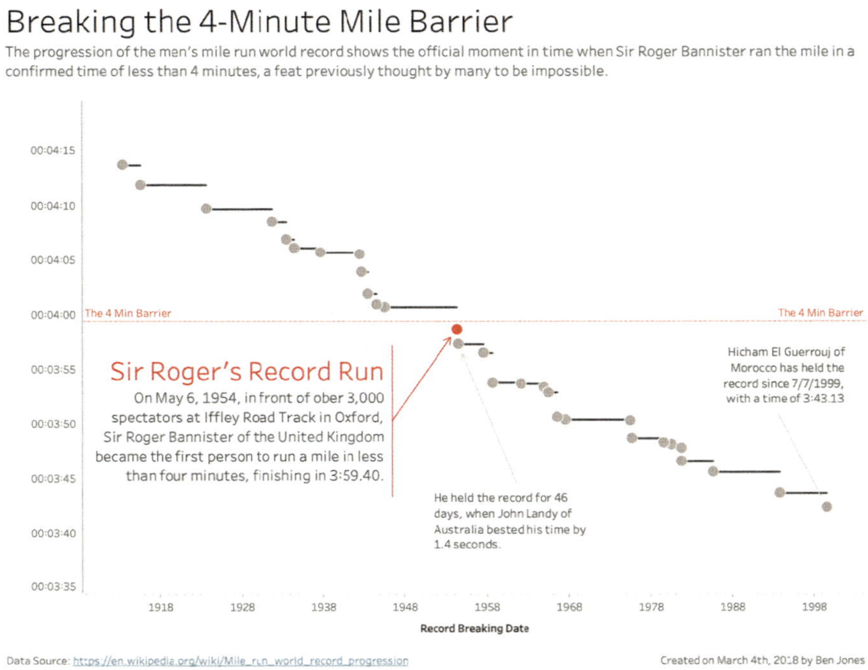

Figure 5.14. Changing world record in the men's mile run shown with jump lines

Note that while both charts follow the "time increases from left-to-right" convention, the mile run jump line chart is an exception to the "up is better" rule of thumb with charts. As athletes run faster and faster, they complete the mile in less and less time, the record time drops, and a new horizontal line begins at a *lower* position on the vertical axis. If it becomes critical to align the vertical direction so that improvements go up, we'd need to instead chart average running speed over the course of the mile, because runners who complete the mile in less time do so by running at a faster rate.

Notice also another subtle detail of this chart. The y-axis doesn't start at 0 minutes and 0 seconds; it starts somewhat arbitrarily at 3 minutes and 35 seconds. For this reason, the x-axis does not have a solid horizontal line at the bottom. Instead, it only shows tick marks for the different gradations of time extending from left to right. This is because the baseline or "floor" of the chart doesn't represent an absolute zero of "no time", or, the moment when the starter's gun goes off to start the race.

Why not? Shouldn't the y-axis start at zero? We've discussed this before relative to bar charts. Should the same rule of thumb apply to line charts? Often it does, but it's safe to say that in practice, the rule is broken more often with lines than with bars. Of course it's impossible to run the mile in 0 minutes and 0 seconds, so the line for the record time will never extend all the way to zero - that's not physically possible. Compare the original version of the jump line chart with the version in Figure 5.15 that *does* have a y-axis that starts at a time of zero.

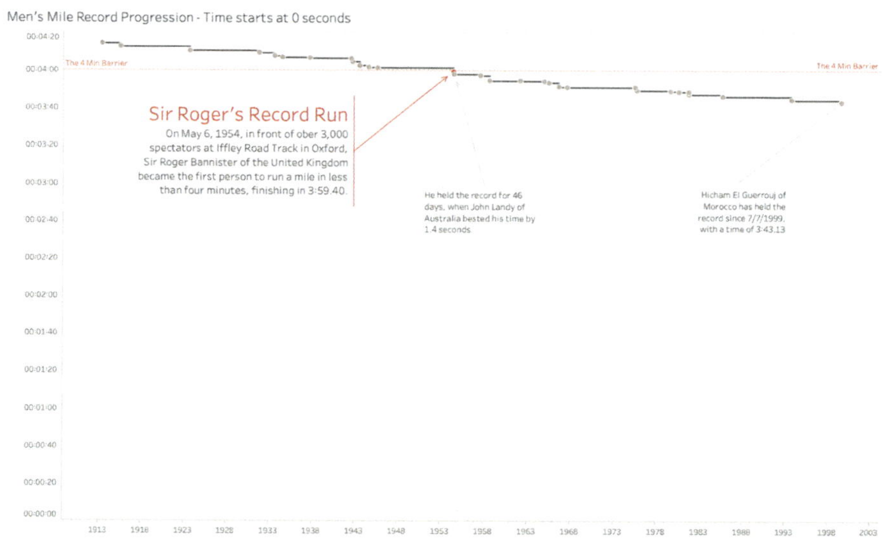

Figure 5.15. The jump line chart with a y-axis that starts at 0:00.

This version of the chart may give us a clearer sense of the *actual* magnitude of each drop in the record time, but it's more difficult to see each individual change, especially when the record is broken multiple times in a relatively short span of time.

So there are trade-offs to these various alternatives that chart-makers and chart-readers alike must consider. As we've considered earlier, there's often no one "right" approach, and the best approach is to evaluate the chart and its suitability for answering our most important questions.

But we can't always depend on our preexisting questions to help us gauge the suitability of a particular chart. Sometimes a chart may give us a misleading answer to our question, and we might not know it. And other times a chart may fail to alert us to an important question that hasn't ever occurred to us—one that's more important than the one we originally conceived.

To illustrate this point, let's consider the New York City rat sighting data set we considered briefly in *Data Literacy Fundamentals*. People in New York City can call 311 to report a rat sighting, and a spreadsheet of each call is compiled and published to the New York City open data portal.

Since each call has its own time-date stamp, we can count the number of calls within various time buckets, or levels of aggregation. How many calls were received in a particular hour, day, week, month, or year? And how have rat sightings changed over the better part of a decade? Four different versions of almost ten full years of data can be found in Figure 5.16. Which is the "best" at revealing the trend?

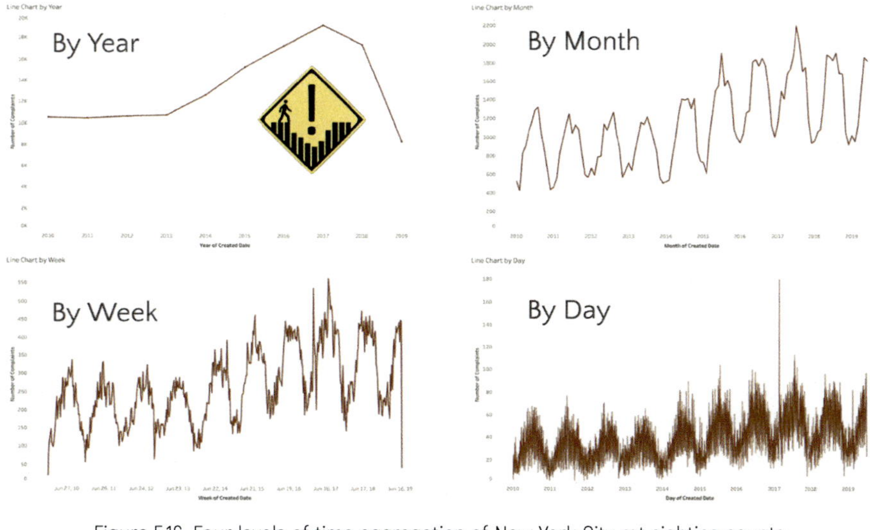

Figure 5.16. Four levels of time aggregation of New York City rat sighting counts

Of course there's no simple answer, but we can certainly compare the benefits and drawbacks of each alternative. The first alternative in the top left, where data has been aggregated by year, does the best job at helping us compare annual increases. We can easily see the increase in calls from 2013 through 2017, and the recent drop-off in calls. This comes with a caveat, however: the precipitous drop in 2019 is misleading, because the data used to make the chart stops at the end of June that year. The dot for the year 2019, therefore, only includes half of a year's worth of data, while the dots for the other years include a full twelve months of calls. We had a question, we used the chart to get an answer, and the answer was wrong.

This leads us to our next Chart Reading Tip:

> **TIP #8:** When data shown on a chart has been grouped by time period, verify whether the time periods are all complete, as some of them may only be partially complete.

When we consider the version of the chart that aggregates counts by month, a brand-new phenomenon emerges: evidence of a seasonal pattern. It seems that reported calls for rat sightings have been much higher in summer months than in winter months. This pattern was nowhere to be found in the line chart showing yearly data because the chosen grouping wasn't granular enough. We can't see which months have higher counts if we lump all twelve months from each year into one single data point.

The weekly version in the bottom left of the figure introduces additional "noise" to the line, and it also presents us with another misleading drop-off at the end of the line. This is due to the fact that June 30, 2019, the final day of data included in the view, happened to be a Sunday. It would appear that whoever or whatever created the chart considered weeks to start on Sunday and end on Saturday. This final week in the line chart, then, includes only one of seven possible days. The other end of the line, the week that started on January 1, 2010—a Friday—only includes two full days of calls, and therefore the very first point in the line is also misleadingly low.

The daily aggregation line chart on the bottom right would seem to be overwhelmingly noisy, as call counts seem to vary significantly at that level of aggregation. Nevertheless, the seasonal pattern is still visible through the noise, and another interesting finding emerges: one day seems to stand out in stark contrast as an outlier: February 15, 2017. We may not have wondered upfront if there was a day with a significantly higher call count than all the others, and the other versions of the chart don't alert us to the fact that such an outlier exists in the data.

This leads us to our next Chart Reading Tip:

> **TIP #9:** Identify what brand new questions the chart prompts in your mind, and consider what further information is needed to answer them.

Notice also that the chart itself, while alerting us to the presence of an outlier, doesn't tell us *why* the outlier happened. To answer the new question or questions that the chart prompts, we often need to get more data or gather additional information. In this case, we might want to ask people in the call center if they remember what happened on that day, or we might want to read the call notes. In this case, simply performing an internet search for "New York rat sightings February 15, 2017" gives us a good clue about what might have happened that day. Give it a try.

Multiple Lines

All of the line charts we've considered so far in this chapter, except for Playfair's original invention, include only a single line for a single series. Technically the jump line chart has many lines, but this set of lines is really just one series broken into segments. But just as bar charts often show more than one bar, line charts often show more than one line. We can view line charts that show separate lines for each of the different levels of a categorical variable in our data set. In this way, we can compare how different items or groups have changed over time.

For example, returning to our global population data set, we can look at how the population of each of the seven regions has changed over time, shown in Figure 5.17.

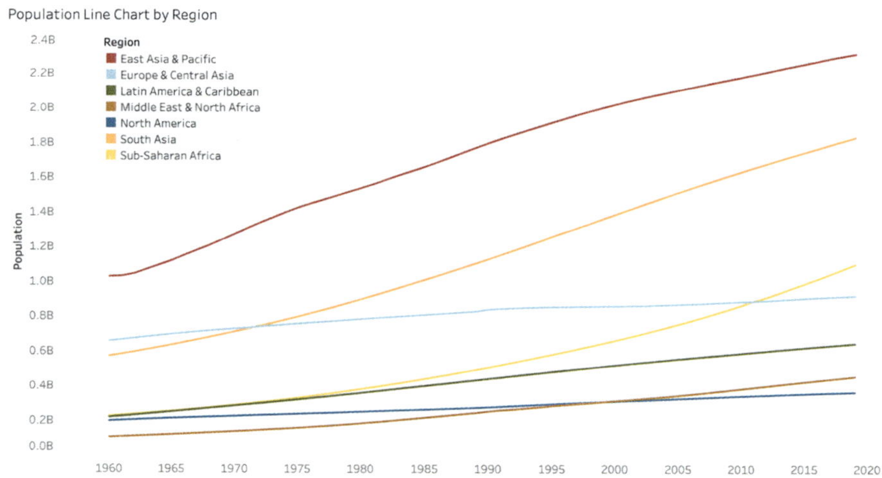

Figure 5.17. A line chart with multiple lines, one for each level of a categorical variable

How interesting that while the overall growth of the world's population forms an almost perfectly straight line since 1960, the changes in regional population have progressed quite differently over that same period of time. We can see that the East Asia and Pacific region is clearly the most populous, and South Asia is second.

We can also see that South Asia surpassed Europe and Central Asia in the early 1970s to claim the second ranked position. And we can see that sub-Saharan Africa also surpassed Europe and Central Asia more recently, and that it seems to be growing at a faster rate than the other regions in recent years.

If we want to more closely consider the relative sizes of the *changes* in population figures for each region, we can use the line charts shown in Figure 5.18.

The version on the left shows us how much the population of each region has changed relative to a single year: 1960. East Asia and Pacific has grown the most in absolute terms since that year, adding over 1.3 billion people to its various countries.

The version of the population change line chart on the right shows us how much the population of each region changed *compared to the previous year*. The previous year's population is subtracted from each year's population figure, and the resulting difference is what gets plotted on the line.

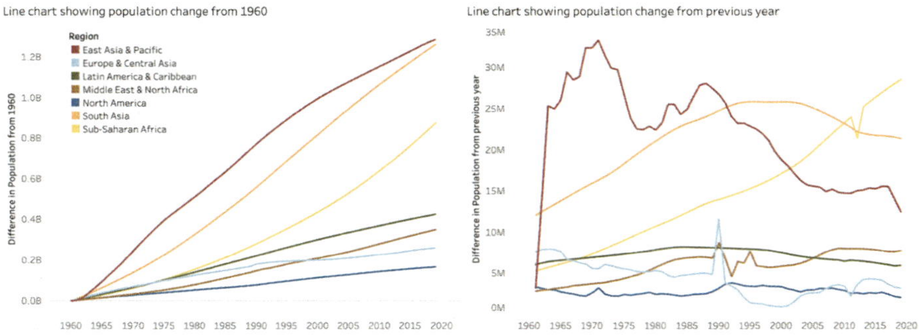

(a) (b)

Figure 5.18. Two line charts that encode change in population

For example, the red line for the East Asia and Pacific region has a value of roughly 19 million in the year 2000, because the population in that region increased from 2.006 billion in 1999 to 2.025 in 2000, for an increase of 19 million. This version of the chart reveals some interesting trends and notches not visible in the other charts, like the spike in the Europe and Central Asia line in 1990 that resulted from Serbia being added to the country list. It also clearly shows that in the past few years, the sub-Saharan Africa region is growing in population by the largest absolute amount.

Instead of comparing *absolute* changes in population (the difference in population between years), we can compare *relative* changes in population by considering percent change, as shown in Figure 5.19.

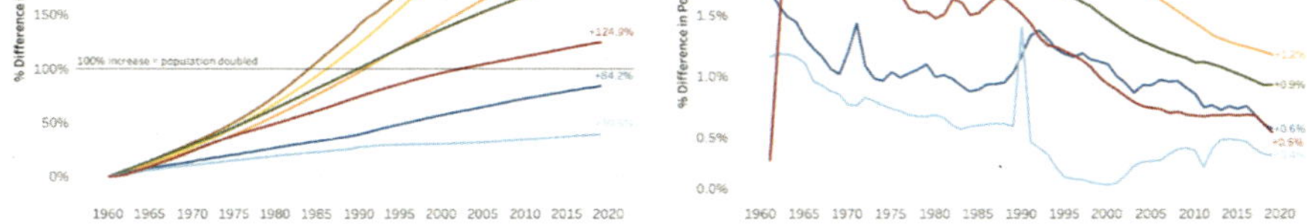

(a) (b)

Figure 5.19. A line chart showing the percent change in population compared to 1960

The version on the left shows percent change in population relative to 1960, and the version on the right shows percent change relative to the previous year. The first shows us that sub-Saharan Africa and the Middle East and North Africa have grown the most in relative terms since 1960: +385.6% and +334.1%, respectively. Note that an increase of 100% is equivalent to a quantity doubling, and an increase of +200% corresponds to tripling, and so on.

The version of the percent change line chart on the right of Figure 5.19 shows us that while most regions have seen drops in year-over-year percent change, sub-Saharan Africa has maintained close to +3% increase each year, with the exception of 2012, when Eritrea saw a mass exodus of its citizens to Europe.

Each of these line charts gives us a different perspective about the way population in each of the seven regions has changed over time. None of these versions of the line chart is inherently better than any other of the versions. Often it can help to look at the data from more than one angle using a different chart or different configuration of the same chart.

This leads us to our next Chart Reading Tip:

TIP #10: When using charts to gain understanding about a topic, try looking at the same quantities and categories using a variety of different encodings and arrangements.

Viewing seven lines on the same chart, each with its own color hue, is a fairly feasible visual and mental task, at least when the lines don't overlap too much. For the most part, we can see each of the lines, we can tell them apart if their hues are distinct enough, and we can learn a great deal from the chart.

Sometimes, though, we encounter charts with dozens or even hundreds of lines. Consider the country population line chart shown in Figure 5.20.

We can clearly see the lines for China and India, and even the lines for the United States and Indonesia. These are the four most populous countries in the world. Below that, though, we can't see much of anything. The lines overlap like pickup sticks lying on the carpet, and we can't use the chart to answer any questions about these countries. Charts with marks that overlap and block each other are said to suffer from occlusion, and they're very difficult to work with.

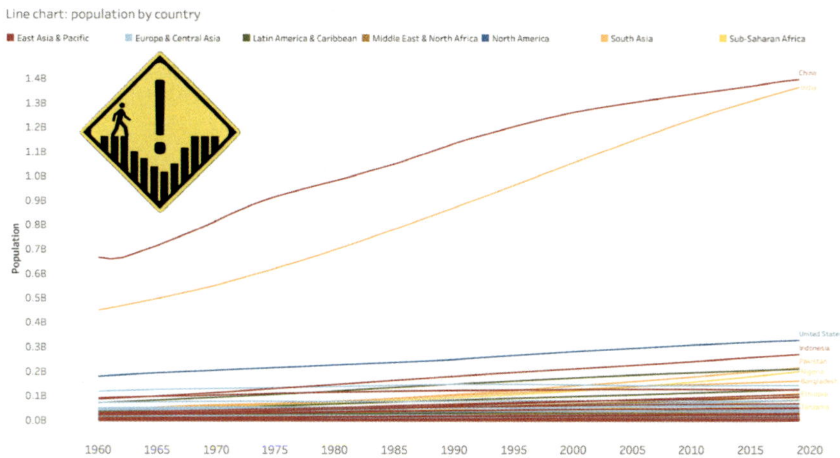

Figure 5.20. A line chart with many lines that occlude each other

In such cases, we can substitute the line chart with a bump chart, a special type of line chart that shows change in *rank* over time as opposed to change in value over time, as shown in Figure 5.21. This chart shows how the ranks of the most populous 25 countries changed since 1960.

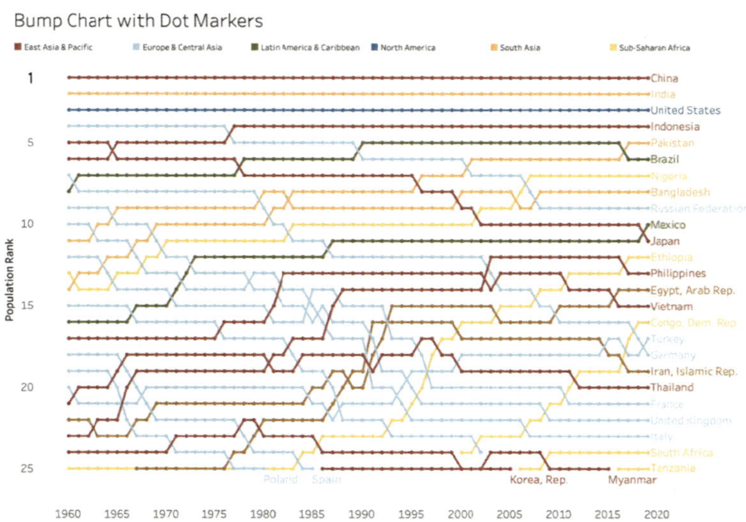

Figure 5.21. A bump chart that shows change in rank over time

There is a trade-off involved with this shift from the line chart to the bump chart. There is always a trade-off. In this case, the trade-off is that while we can clearly see how the ranks have changed, it is impossible to see just how far apart the countries are in actual population. Remember how far above the rest China and India are on the standard line chart? This fact is lost in the bump cart. The line for India is as close to the line for the third-ranked country, the United States, as the line for the 24th-ranked country is to the 25th-ranked country.

The Area Chart

Recall that there are three basic mark types: points (zero dimensional, or 0D), lines (1D), and areas (2D). The fourth mark type, volumes (3D), is far less common since we tend to read charts on two-dimensional surfaces like screens or pages, and depth and volume are very difficult to gauge when they are projected onto a flat surface. It remains to be seen whether virtual or augmented reality (VR/AR) will bring charts with 3D marks and encodings into more prominence in the future as these technologies continue to evolve.

When we began our consideration of change over time, we started by looking at dots (0D) on a Cartesian coordinate system with time on the horizontal axis. We then moved to lines (1D) by connecting the dots in various ways: straight lines, step lines, and jump lines. We can continue this progression and move from lines to areas (2D) by filling in the gap between the line and its horizontal axis, resulting in the area chart shown on the right side of Figure 5.22.

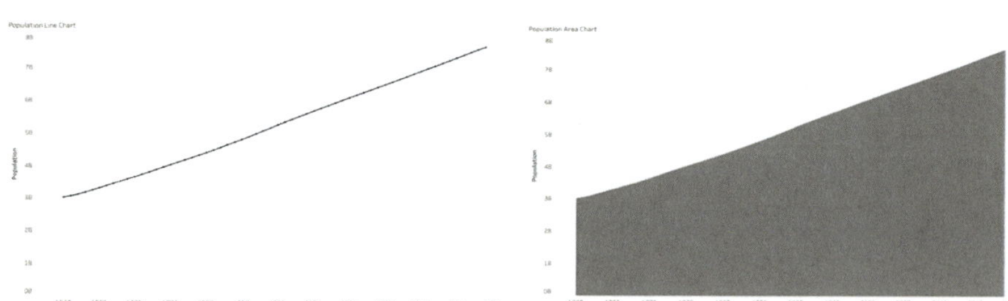

(a) (b)

Figure 5.22. From a line chart (a) to an area chart (b)

In this simplest version, the area chart involves little more than a stylistic shift from the line chart. Instead of focusing on the line, we're looking at a shaded region to make judgments about the changing population of the world.

Note that the y-axis starts at a population of zero, so the distance from the line to the horizontal axis is proportional to the population. That's good because the height of the shape maps directly to the quantity being encoded. If the y-axis started at 2.5 billion, the height of the shaded region would be misleadingly small, similar to the way truncated bars can give a skewed sense of the true proportions.

What about the width of the shaded region, though? Population is a quantitative variable that can be summed within a year, but it would be unusual to add populations of different years together. A country's population in one year includes many of the exact same people that lived there the year before. Adding last year's population of Germany to this year's population of Germany, for example, doesn't yield a very meaningful sum. This is different from, say, a company's sales figures, which are additive from year to year as a company continues making money as time goes on. Last year's sales plus this year's sales tells us how much the company sold in the past two years—a meaningful sum.

For this reason, an area chart might not be an ideal way to show changes in population over time. The single shaded region under the line expresses the notion that the populations across the years can all be added together into a meaningful grand total, but they can't.

A further challenge with area charts is that it can be difficult to see how *multiple* areas have changed over time, especially as the number of areas increases beyond just two or three. The areas overlap and tend to "occlude" each other. Even making the areas semitransparent won't necessarily fix the problem entirely. Compare the line chart on the left of Figure 5.23 with an area chart on the right that shows the exact same seven time series as superimposed, semitransparent areas. It's nearly impossible to distinguish the different trends, especially toward the bottom of the chart.

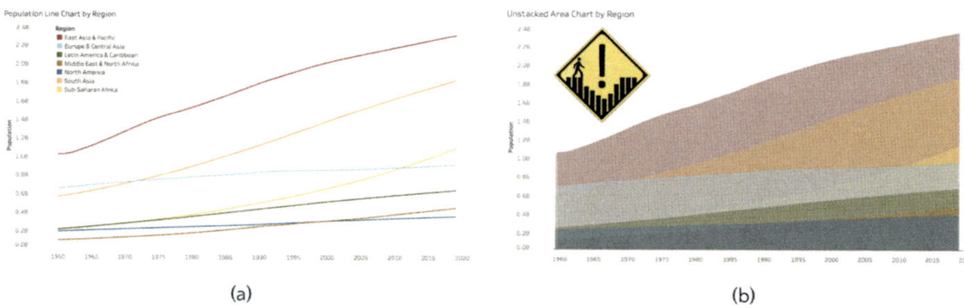

Figure 5.23. A multiple line chart (a) and a multiple area chart (b) using the same data

For this reason, area charts with multiple levels of a category tend to show stacked areas instead of unstacked, or superimposed, areas, similar to the way the chart would look if it encoded populations with stacked bars, as shown in Figure 5.24.

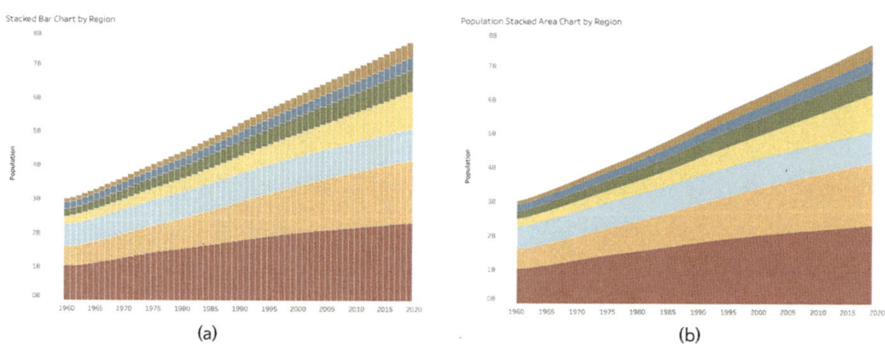

Figure 5.24. From a stacked bar chart (a) to a stacked area chart (b)

Of course this only works if the values can be added *within* a particular time period. That's true of population; we can add the populations of each of the regions together in any particular year to get the population of the entire world that year. Or we can add the population of, say, the United States with the population of Canada to get the total population of those two countries put together.

That isn't the case, though, with non-summable variables like, say, life expectancy or graduation rate. If we add the life expectancy of one country to the life expectancy of another country, we don't get a meaningful total. The life expectancy of someone born in the United States in 2017 was 78.54 years. The life expectancy of someone born in Canada that same year was 81.95 years. If we add those two values together, we get a value of 160.49, but this number doesn't mean anything.

Because of this, we wouldn't want to stack these non-summable variables, whether we're using a stacked bar chart or a stacked area chart. We wouldn't even want to see them as slices in a pie chart, because they aren't parts that can be added together to form a meaningful whole.

When we do have part-to-whole data, though, such as with the populations of the different regions of the world, we can see the change in percent of total for each of the parts using a stacked area chart, as in Figure 5.25b.

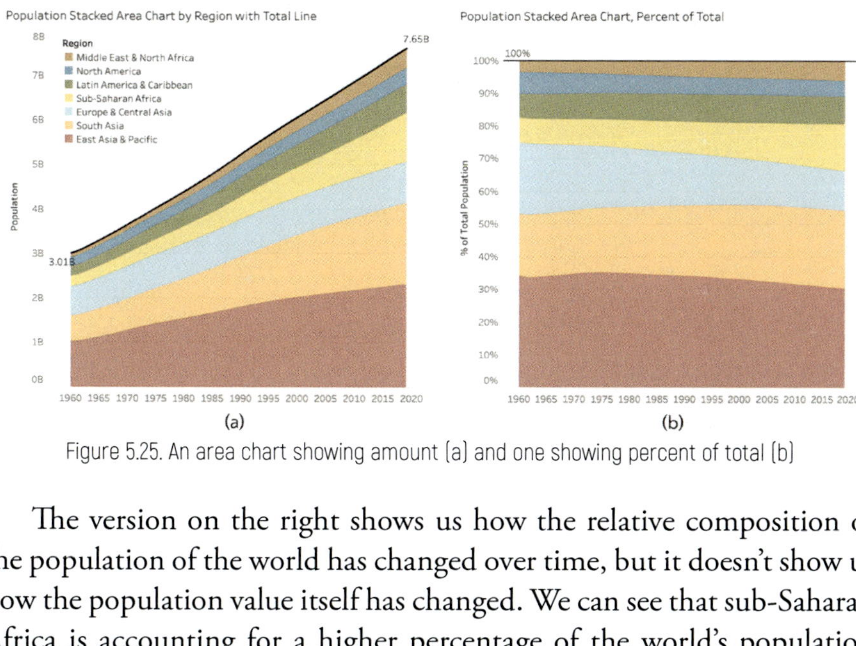

Figure 5.25. An area chart showing amount (a) and one showing percent of total (b)

The version on the right shows us how the relative composition of the population of the world has changed over time, but it doesn't show us how the population value itself has changed. We can see that sub-Saharan Africa is accounting for a higher percentage of the world's population than it did in the mid-twentieth century, and that the opposite is true of Europe and Central Asia.

Note, also, that both the stacked area chart and the percent of total area chart in Figure 5.25 also include a dark line at the top, which shows how the total population changes over time (a) and the fact that the areas all add up to 100% for each year (b). This is an advantage that area charts have over the line chart—a clear depiction of the total amount.

The line charts we considered in the previous section did not show the total amount; they showed only the lines for each of the different region groups. We can modify the line chart, though, to add a line for the total amount, as in Figure 5.26. When we do that, the individual lines for the regions get pushed to the bottom, making it difficult to compare their individual trends over time. Compare the line chart with black total line on the left to the stacked area chart with black total line on the right and think about what kinds of questions are easier to answer with each one.

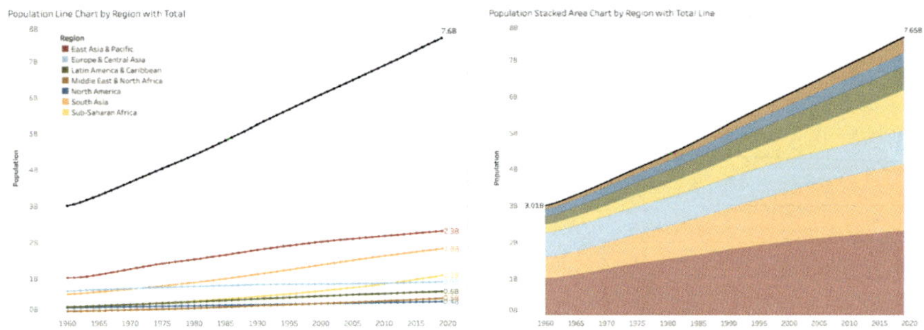

(a) (b)

Figure 5.26. A line chart with a total line (a) compared to an area chart with a total line (b)

One potentially misleading aspect of the stacked area chart is that the person reading it can sometimes make the mistake of assuming that the individual areas are behind each other as opposed to stacked on top of one another. When there is a lot of variation, noise, or even a seasonal pattern, that might lead them to jump to the wrong conclusion.

For example, when we consider building permit applications in the city of Seattle, we can see how the total number of permit applications has changed over time in the stacked area chart in Figure 5.27. This area chart breaks down the total number of applications into seven different permit classes (single-family / duplex homes, commercial buildings, multifamily apartment complexes, etc.), each with their own regions with distinct color hue.

The wave-like shape of the overall shaded region gives us a clue that there is some pattern to the data. The pattern is referred to as seasonality—a regular, predictable pattern related to the months of the year. The peaks of the line at the top are all summer months, and the troughs or valleys of the shaded region are all winter months. It would seem that the City of Seattle receives more applications for building permits when it's warm than when it's cold.

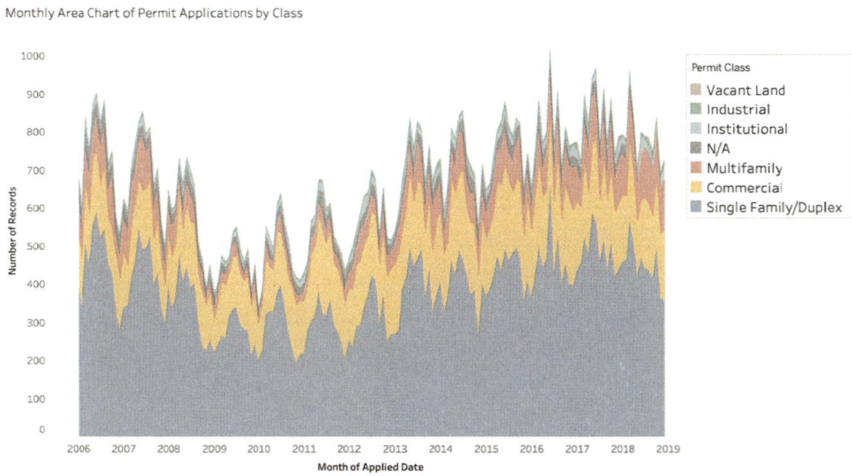

Figure 5.27. A stacked area chart showing seasonality of permit applications in Seattle

When you look carefully at this chart, though, can you tell which of the permit classes exhibit seasonality? Is only one of them seasonal, or do each of them exhibit this pattern? When asked this question, many people reply that they are all individually seasonal. If we reorder the areas, though, and place the blue region for Single Family/Duplex at the top of the stack instead of at the bottom, as in Figure 5.28, we can see that only the blue category exhibits seasonality, and the others don't have that same wave-like shape to them after all.

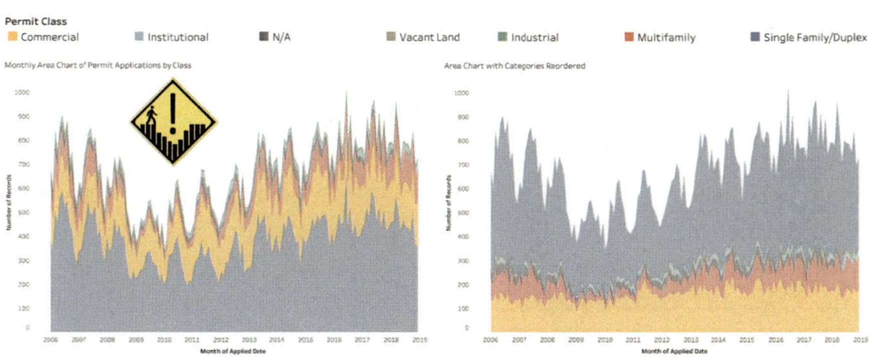

Figure 5.28. Two versions of the same stacked area chart differing in the order of colors

The area charts we have considered so far have all made use of the Cartesian coordinate system. The horizontal axes have all encoded time, progressing forward from left to right, and the vertical axes have all encoded quantity or amount, increasing from the bottom to the top. But just as it was possible to convert part-to-whole encodings from rectangular to polar by creating a pie chart, it's also possible to show change over time using the polar coordinate system.

That's just what Florence Nightingale did in 1858 when she created her famous polar area diagram, sometimes referred to as the coxcomb plot. Nightingale and her charts (see Figure 5.29) are the inspiration for the cover of this book. She was known as the "Lady with the Lamp" because, as a nurse in the English army during the Crimean War, she would carry her lamp as she did the rounds in the hospital at night, checking on soldiers.

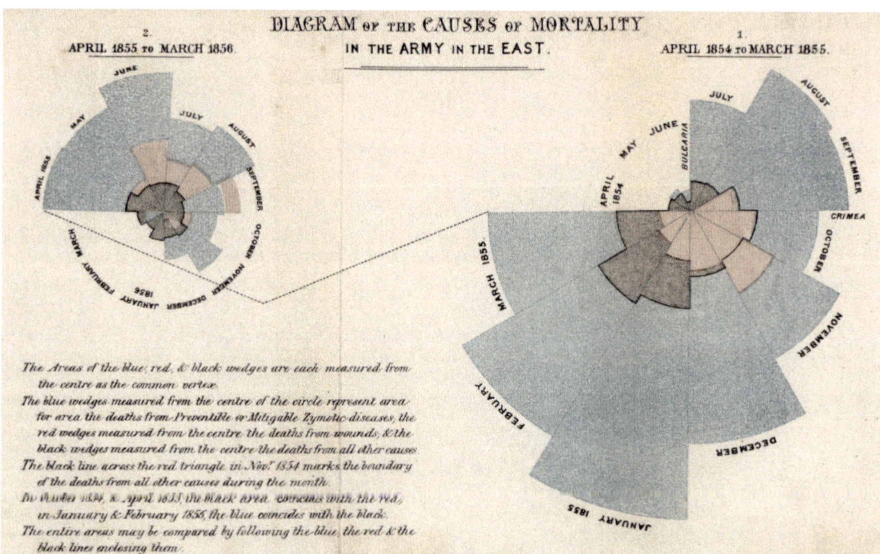

Figure 5.29. Florence Nightingale's coxcomb plot (Public domain, via Wikimedia Commons)

The two polar area charts in the figure are composed of twelve wedges each. A wedge corresponds to a month of the year, and the area of each wedge is proportional to the number of British soldiers who died in field hospitals from April 1854 to March 1856. The colored sections of each wedge correspond to the cause of death—blue sections for preventable causes such as disease, red sections for wounds suffered in battle, and black sections for all other causes.

Notice the surge in deaths in the colder winter months that are primarily due to disease, as sick and wounded soldiers huddled together in hospitals with little to no sanitation. Notice also the dramatic reduction in such deaths the following year, indicated by the smaller diagram on the left. What caused this drastic improvement?

Nightingale, a nurse by profession but a statistician by training, was able to collect this data, publish these creative charts, and return to England to make the case to Queen Victoria to institute sweeping sanitation reform in these field hospitals.

Her example of using charts that show change over time in order to make a strong and convincing case for meaningful policy change is what using data is all about.

Other Ways of Showing Change Over Time

In our survey of charts that encode change over time, we've considered line charts, bump charts, and area charts. We can also use the other charts we've learned about so far, and simply place versions of the chart from different time periods next to each other, or juxtaposed.

For example, if we wanted to show how the breakdown of the population of the world into its different regions was different in 2019 compared to 1960, we can consider two pie charts side by side, with the older version on the left, as shown in Figure 5.30. In this way we can see how the part-to-whole relationship was different in one year as compared to another. Notice also how the overall areas of the two pies are different. They encode the overall population of the world, which as we know increased from 3 billion in 1960 to 7.64 billion in 2019.

There are many more ways to show change over time with data, including practical ways like the Gantt chart for project tracking and artistic ways like the steam graph for a more fluid feel. Most of these alternate ways of showing change over time also use the left-to-right convention, but not always.

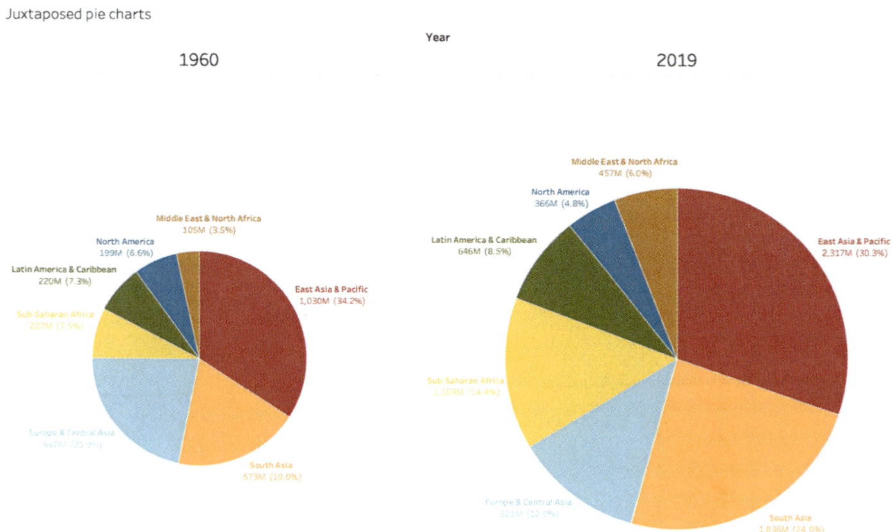

Figure 5.30. Two juxtaposed pie charts showing part-to-whole along with change over time

The bar chart race, for example, doesn't map time to space at all, but instead maps time to time, animating the bars and giving a reader a feel for how the time series actually played out. The advantage of this technique is that it can sometimes evoke a sense of awe. The disadvantage is that the reader can't see the entire time series all at once, making it difficult to use for analytical purposes.

As we're learning, each chart type has its pros and cons.

Being able to understand how our environment has changed over time is such an important skill for each of us to master. We come across the charts we've reviewed in this lesson almost every single day, whether we're tracking the stock market, the weather, or our bank account balance.

FURTHER LEARNING

Mind in Motion by Barbara Tversky (Basic Books, 2019): https://www.basicbooks.com/titles/barbara-tversky/mind-in-motion/9780465093076/

Tversky, B. et al. "Cross-cultural and developmental trends in graphic productions." Cognitive Psychology 23 (1991): 515-557. https://www.sciencedirect.com/science/article/pii/0010028591900059

Santiago, J., Lupáñez, J., Pérez, E. et al. Time (also) flies from left to right. Psychonomic Bulletin & Review 14, 512–516 (2007). https://doi.org/10.3758/BF03194099

Chapter 6

UNDERSTANDING VARIATION

"Variety's the very spice of life, that gives it all its flavour."
—William Cowper, *The Task* (1785)

Thankfully, we don't live in a bland, uniform world. We're incredibly different from each other, and there's variety in everything around us. It's the stuff of poetry. As the epigraph states, this is what makes life so interesting.

Luckily, our data is no different. The counts we tally and the measurements we collect result in data sets that can contain incredibly rich variation. Like a landscape that isn't perfectly flat, but rather filled with mountains and valleys, our data contains high points, low points, and everything in between.

We know how to see and appreciate the differences in the world around us; we just open our eyes and look around. The good news is that we can use our eyes, combined with the encodings we've learned about, to see how quantities in our data vary, too. It can be a thrill to discover the texture our data actually contains, and to ponder what it means about our surroundings, both near and far.

But we'll need more than just charts to get a firm grasp of this critical aspect of our data; we'll also need to understand a handful of statistical measures that capture and convey two important aspects of our quantitative variables: central tendency and dispersion. Let's cover these concepts first.

Central Tendency and Dispersion

We've already looked closely at how to compare quantities for different categorical variables, like population by region. The bar chart proved to be a useful way to encode position, length, and area to give our minds three ways to judge relative amounts. And we've also considered different ways to see how a total amount can be broken down into its constituent parts with the stacked bar chart, the pie chart, or the treemap.

But we have yet to consider how a single quantitative variable—any measurement, count, or reading taken multiple times—*varies* from reading to reading in a given data set. If we collect readings about different items in a group, often the resulting data set will include more than one value in it. Sometimes a list of readings is completely uniform, meaning it contains the exact same value over and over. More often, though, a single quantitative variable will include values that differ from one another to some degree. Diversity is the rule, uniformity the exception.

This variation that exists in even a simple list of numbers causes two questions to very naturally occur to us: first, what's a *typical* value in the set? Second, how much do the values tend to *differ* from one another?

The first question about what's typical relates to what's called central tendency. Simply, measures of central tendency tell us where the center of the values can be found. The second question relates to the notion of dispersion, or variability. Measures of dispersion help us understand the degree of spread in the values.

If you find the phrase "statistical measures" to be intimidating, don't worry, these concepts aren't rocket science. We'll get through them together so that you have a firm grasp of them. There are three main measures for each of the two groups of questions that we'll consider here: three measures of central tendency and three measures of dispersion. If you've already gone through the Data Literacy Fundamentals book or course, this will be a review for you, but we've mixed things up to consider the concepts with a new data set.

Let's consider these two questions one by one using the following data set that hopefully you'll find engaging. First we'll consider the measures themselves. Then, we'll discuss how our charts can best convey them to us.

To begin, imagine seven people standing on a bridge. They work together and they're on a bold mission, exploring places where no one has gone before. We'll call them the Bridge Team. Their heights are listed in Figure 6.1.

Name	Gender	Department	Height	Height in inches
Patrick	male	Command	5'10"	70
Jonathan	male	Command	6'3"	75
Marina	female	Science / Medical	5'5"	65
Denise	female	Operations	5'8"	68
Brent	male	Operations	5'11"	71
Gates	female	Science / Medical	5'8"	68
LeVar	male	Operations	5'7"	67

Figure 6.1. A table of the heights of seven people standing on a bridge

Of course, we know from what we've learned already that we could visualize individual team member heights as vertical bars (columns), or we could even visualize the percent of the total height contributed by each department with a pie chart as shown in Figure 6.2.

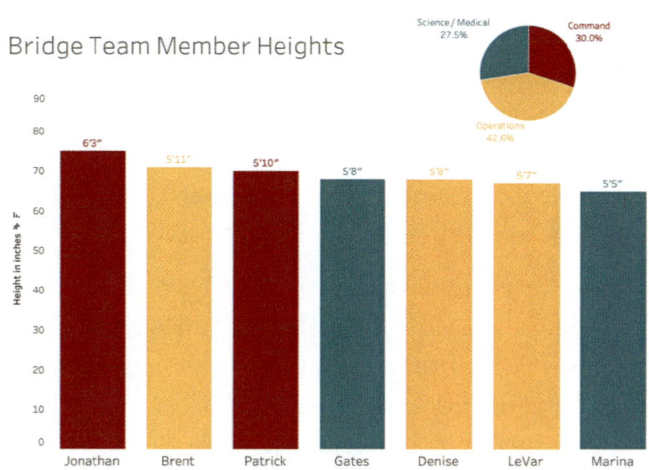

Figure 6.2. Two visualizations showing bridge team member heights

And while the bars are useful for helping us see and compare the heights at a glance, the pie chart seems strange because we don't usually think of adding heights like this. People don't stand on each other's heads very often, so while summing heights is technically allowable, it isn't exactly common or useful. Instead, our minds want to know if there's a typical height for the group, and how much individuals vary from that typical value. Enter central tendency and dispersion.

1. Measures of Central Tendency

A quick glance at the table reveals to us right away that their heights are not all the same. Some of the explorers are shorter, and others are taller. That's no big surprise. But is there a "typical" height, even for this small list of people? How would we answer that? There are three main ways. They all start with the letter M, which makes them easier to remember.

A. **MODE:** The simplest way to answer the question is to find the value that occurs the most often. This is called the mode. Two people, Denise and Gates, have a height of 5'8", or 68 inches. None of the other height readings appears more than once in the list, so we have our mode: 68, or 5'8".

B. **MEAN:** The most common way to find a central point in the values is to find the arithmetic average, or mean. We come across averages every single day. They're everywhere. To find the mean, we simply add all the values in the set together, and then we divide that sum by the total number of values in the set. The sum of the height readings is 484, and if we divide that by 7, the number of individual readings, we get 69.14, or a little greater than 5'9".

C. **MEDIAN:** Another way to define a central point is to first sort the list, and then find the value in the middle. This value separates the greater, or larger half of the readings from the lesser, or smaller half. It's also known as the 50th percentile, because it's greater than 50%, or half, of the other readings. Doing so with this list of seven numbers gives us a median of 68 inches.

Figure 6.3 shows all three ways of finding the central point. Note that they are not all the same. The mean, median, and mode are three different ways of defining central tendency, and it's rare that they all give us the exact same result. Like chart types, not one of them is "right." Considering them together can give us a better understanding of our data.

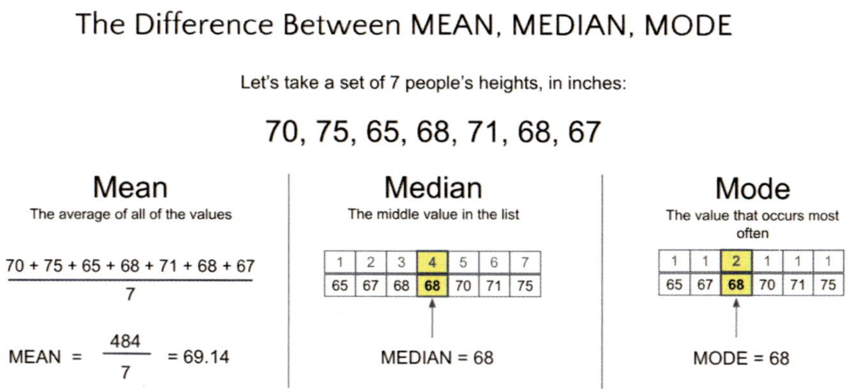

Figure 6.3. A simple example of measures of central tendency

Now that we have a handle on various ways to measure the central value, we next turn to measures of dispersion: what's the spread of the data; how much do the individual readings vary?

2. Measures of Dispersion

As with measures of central tendency, there are three ways to measure dispersion.

A. **RANGE:** The simplest way to measure and communicate the spread of the data is to calculate the range, which is just the maximum value minus the minimum value. In this case, the maximum value is 75 inches and the minimum value is 65 inches, so the range is 10 inches.

B. **VARIANCE,** σ^2 ("sigma squared") or S^2: The variance depends on whether the data set makes up an entire population, or just a sample of the population. In other words, are these seven people the only people on the team, or have they been randomly selected from a much larger group? Let's start by assuming they have a much larger team on a ship in the area, and we'll compute the variance and standard deviation for a sample.

Computing the variance can be best described using a five-step process. None of these steps are complicated at all, but the fact that there are five steps can make it a little tricky for many. See if you can follow along, step by step. Refer to Figure 6.4 to see the steps put into practice.

Calculating the Variance in 5 Easy Steps

		Step 1	Step 2	Step 3
Name	**x (height)**	**μ (average)**	**x - μ**	**$(x - \mu)^2$**
Patrick	70	69.14	70 - 69.14 = 0.86	$(0.86)^2 = 0.73$
Jonathan	75	69.14	75 - 69.14 = 5.86	$(5.86)^2 = 34.31$
Marina	65	69.14	65 - 69.14 = -4.14	$(-41.4)^2 = 17.16$
Denise	68	69.14	68 - 69.14 = -1.14	$(-1.14)^2 = 1.31$
Brent	71	69.14	71 - 69.14 = 1.86	$(1.86)^2 = 3.45$
Gates	68	69.14	68 - 69.14 = -1.14	$(-1.14)^2 = 1.31$
LeVar	67	69.14	67 - 69.14 = -2.14	$(-2.14)^2 = 4.59$

Step 4: 0.73 + 34.31 + 17.16 + 1.31 + 3.45 + 1.31 + 4.59 = 62.86
Step 5: 62.86 / (N - 1) = 62.86 / 6 = **10.48**

VARIANCE = 10.48

Figure 6.4. How to calculate the variance of a sample

1. First, we take the mean or average value, μ (the Greek letter mu, pronounced "myu") of the set. We've already calculated the mean from our measures of central tendency, μ = 69.14, so we put that value in the Step 1 column.
2. Second, we subtract the mean, μ, from each value, x, to get x - μ. Results for this step are in the Step 2 column.
3. Third, we square the values we got in the previous step, or, in other words, we multiply them by themselves. The result, $(x - \mu)^2$, can be found in the Step 3 column.
4. Fourth, we add up all of the values we got in Step 3. This gives us 62.86, and is shown beneath the table in the row labeled Step 4.

5. Fifth, we divide the number we got in Step 4, 62.86, by the number of values in the set, N, minus 1: N—1, or 7—1, or 6. This gives us 62.86 / 6 = **10.48**, and that's our variance!

   The fifth and final step in the process is slightly different if we're dealing with all of the members in a population rather than just a sample of the population. In such cases, instead of dividing by N—1, we instead divide by N.

   The good news is that you'll very rarely have to do this calculation by hand. Most software packages let you compute this statistic directly, and even Excel lets you enter the function "=VAR(first.cell:last.cell)."

Interestingly, the variance isn't used nearly as much as the next measure of dispersion, the standard deviation:

C. **STANDARD DEVIATION**, σ ("sigma") or S, sometimes abbreviated SD: The standard deviation is used quite often in statistics, business, manufacturing, and many other contexts. It's calculated by simply taking the square root of the variance. In our simple example, $\sqrt{10.48} = 3.24$.

   But what does this value actually tell us? Well, since standard deviation is a measure of the spread, a low standard deviation means that the individual values are located very close to the average, or mean. On the other hand, a large standard deviation means that the values are far away from the mean.

Knowing how to calculate these important measures, we can include them in the charts we've already covered in this book. Going back to our World Bank country data set, we can use our encoding channels to *see* the difference between the average country population of the different regions.

Instead of encoding bar length with the sum of the population of countries in a region, we can instead use the average population of each region to encode this same channel, as shown on the left of Figure 6.5. Or we can enhance a dot plot of country populations by region by adding a triangle shape or reference lines for the average values, as shown in the chart on the right of Figure 6.5.

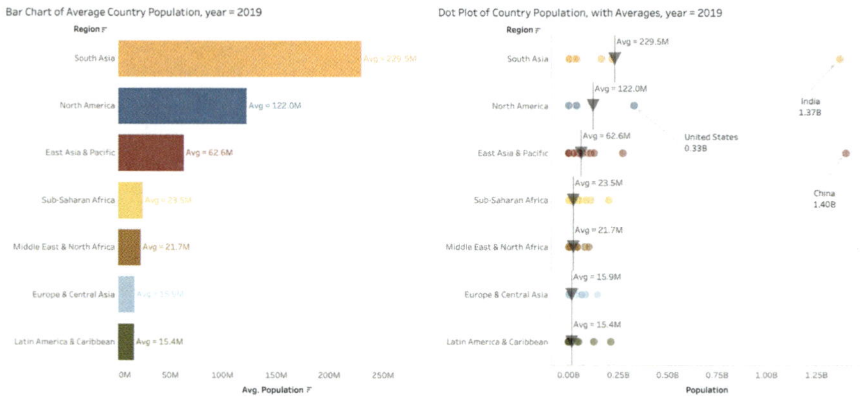

(a) (b)

Figure 6.5. Encoding visual channels (bar length and dot position) with average values

In this way, we're using the average value instead of the sum of the values as the method of aggregation in our charts. We could similarly use the median or the standard deviation or any of the other measures of central tendency and dispersion to aggregate the individual values and encode the visual channels in the chart. The important question to answer is: which method of aggregation will be the most helpful?

And that gives us our next Chart Reading Tip of the book!

TIP #11: Consider how the data in the chart has been aggregated (sum, average, median, etc.), and whether it's suitable to answer your most important questions.

We need to be very careful with some methods of aggregation. They're not all valid for all variables. Consider life expectancy, for example. As we've already considered, each country has an estimated life expectancy for a given year. That's how long people born in that country in a certain year are expected to live. And as we discussed in the previous chapter, it wouldn't be meaningful to add these values together or stack them in a chart because they aren't summable.

Similarly, the average life expectancy between two or more countries doesn't really give us a meaningful value. Why not? Because the countries may be dramatically different in size, so we'd get a very skewed result if they all carried the same weight in the calculation.

Take, for example, China (1.39 billion people, life expectancy = 76.47 years in 2017) and Somalia (14.6 million people, life expectancy = 56.71 years in 2017). So what's the average life expectancy for these two countries, then?

If we just take the average of the two values, we get (76.47 + 56.71) / 2 = 66.59. But this isn't valid, because China is so much more populous than Somalia. In cases like these, we compute a weighted average as the default form of aggregation, rather than a simple arithmetic average. China accounts for 99% of the combined population of these two countries, so we multiply its life expectancy by 0.99, and we add it to the life expectancy of Somalia times 0.01 (since it accounts for 1% of the combined population). This weighted average gives us a combined life expectancy of 76.2 years.

The Histogram

The measures of central tendency we learned in the previous section are helpful, but often we want to be able to actually *see* how quantities are distributed. The statistical measures alone won't give us a clear picture of the shape of our data. The good news is that there are chart types that will.

One such chart type is the histogram. It looks a lot like a bar chart, and it's often mistaken for one, but they're different. The bar chart, as we've seen, lets us compare quantities that have been broken out into the levels of a categorical variable, such as population by region. We see a form of this on the left side of Figure 6.6.

The basic histogram is different from the bar chart because it doesn't use any categorical variables at all. Instead, it shows how individual values of a single quantitative variable can be distributed into adjacent "bins," or "buckets," that span the range of the data. This gives the reader a sense of how the values are distributed. As you can see by comparing the left side (bar, or column chart) with the right side (histogram) of Figure 6.6, the adjacent bars of the histogram touch each other, with no gap in between, while bar charts often have larger gaps between the bars:

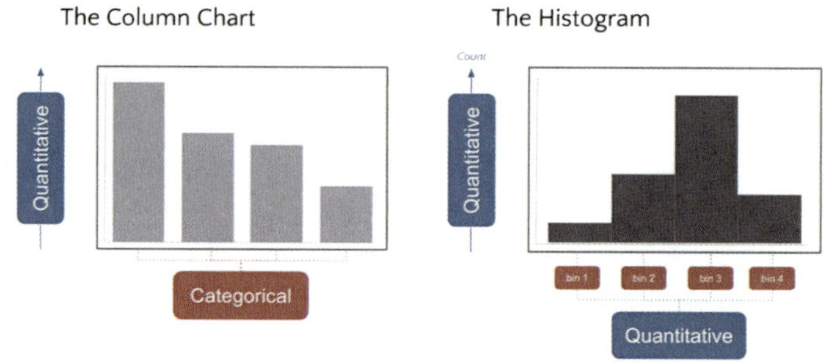

Figure 6.6. The difference between a bar chart and a histogram

Let's consider an example. If we wanted to see the shape of the distribution of the seven heights of the "Bridge Team" members we considered earlier, we can place them into bins of 3 inches, as shown in Figure 6.7. We can see that there's one person, Marina, in the first bin, which is for people on the list who are greater than or equal to 63 inches tall, and less than 66 inches tall. She's 65 inches tall, so she belongs in that first bin. The next bin, 66 to 69 inches, has three members in it: Denise, LeVar, and Gates. And so on. The upper limit of each bin is noninclusive, meaning that a person on the line would fall into the higher bin. This is why Jonathan, who is 75 inches tall, falls into the 75- to 78-inch bin instead of in the 72- to 75-inch bin, which is empty.

The version of the chart at the bottom of Figure 6.7 shows the standard histogram, with rectangles whose heights are proportional to the count of team members in each bin.

Let's consider another example using real data. We'll return to our World Bank country data set, and we'll see if the histogram can reveal to us the shape of the distribution of the life expectancy variable.

In 2017, the country with the highest life expectancy was Hong Kong. Those born there had a life expectancy of 84.68 years. This value represents the maximum in the set of values for that year. On the other extreme, the Central African Republic had a life expectancy of 52.24 years in 2017, the minimum in the set. So the range was 32.44 years (84.68—52.24).

To create a histogram of life expectancy, let's start by grouping countries into bins with a size of 5 years, starting with a bin from 50 to 55 years, followed by a bin from 55 to 60 years, and so on, up to a final bin of 80 to 85 years. Each country is placed into the bin that corresponds with its life expectancy, and we stack short rectangles on top of each other as shown in Figure 6.8, one for each country in its respective bin.

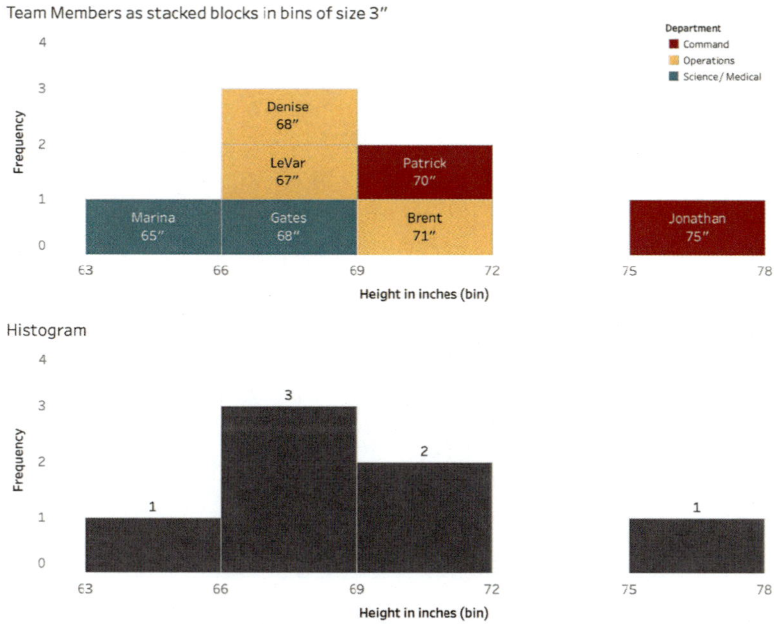

Figure 6.7. How the histogram is formed: stacked

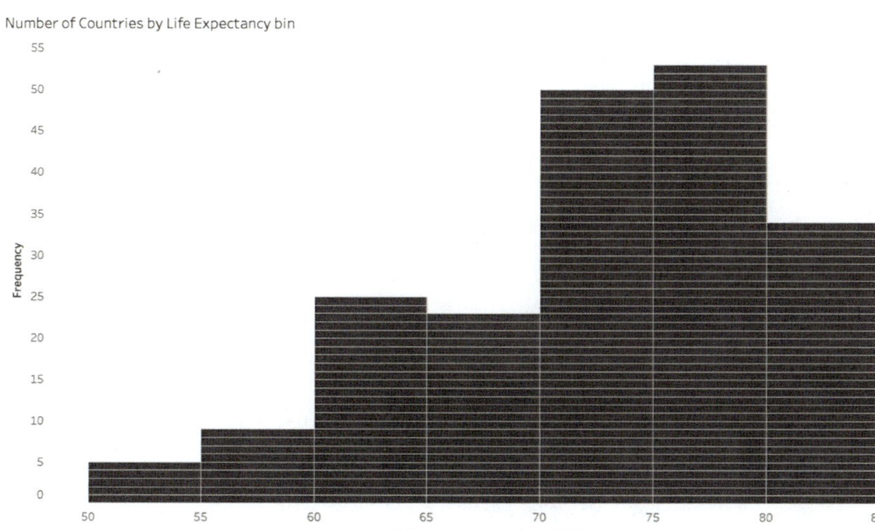

Figure 6.8 Stacking country disks on top of each other in life expectancy "bins"

Thus we can see that there are five countries in the first bin, with life expectancies between 50 and 55 years, nine countries in the next bin, 25 in next, and so on. The standard histogram, shown in Figure 6.9, simply removes the lines that define each country's rectangle, blending them together into a single rectangle above each defined bin.

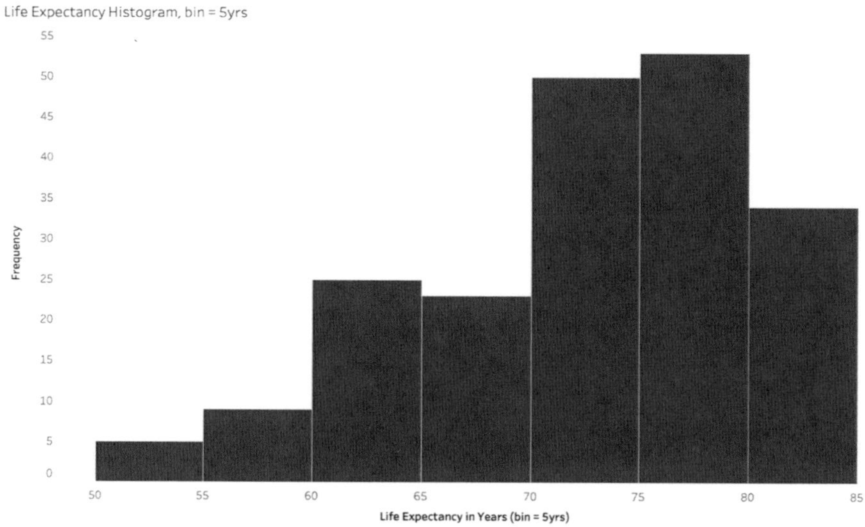

Figure 6.9. Histogram of country's life expectancies in 2017, bin size = 5 years

The bin sizes of a histogram can be modified, and, interestingly, the shape of the exact same distribution can look very different when different bin sizes are used. The histogram in Figure 6.9 uses a bin size of 5 years, but there's nothing that says it *has to be* that size. Figure 6.10 shows four different histograms using this same country life expectancy data set. The bin sizes vary from 1 year (top left), 5 years (top right), 10 years (bottom left) and 25 years (bottom right).

Which one is the best? It seems pretty clear that the histogram that uses the largest bin size of 25 years isn't very helpful or interesting. It certainly doesn't reveal much about the underlying distribution. One the other end of the extreme, the histogram that uses a bin size of 1 year has a lot more "noise" in the shape, but perhaps too much noise.

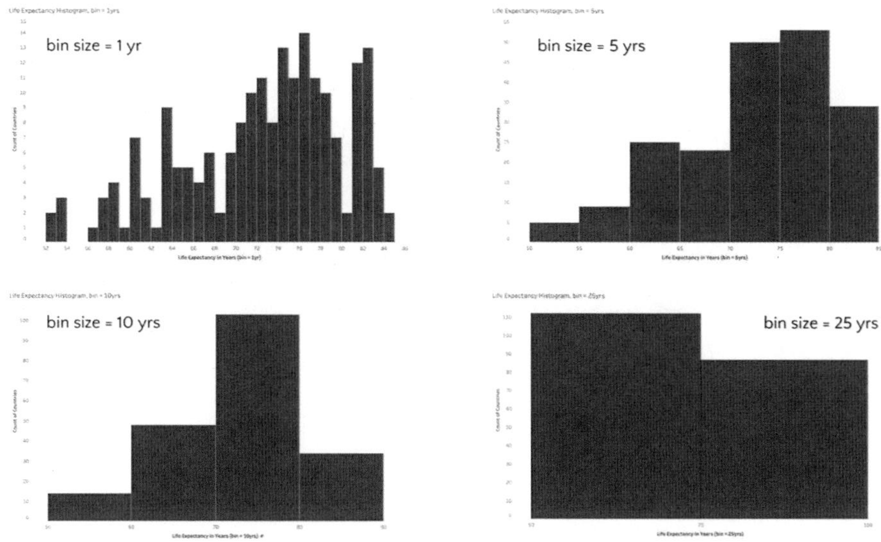

Figure 6.10. Histograms of different bin sizes made with the same data set

Our original histogram that uses a bin size of 5 years seems like it might be the "Goldilocks" option that's just right. It shows us that many countries are to be found in the 70- to 75-year bin and, to even more so, in the 75- to 80-year bin. It also reveals a longer "tail" of bins with lower life expectancies to the left. This isn't quite as clear in the other versions.

In general, it's useful to try looking at the same histogram with a number of different bin sizes. It's hard to predict which one will be the most helpful, so consider bin size, and what different settings of this parameter look like.

While basic histograms do differ from bar charts in that they don't involve any categorical variables, that's not to say that they *can't* incorporate categorical variables. We could consider, for example, how the distributions of life expectancies in 2017 were different for different income groups, or regions of the world. See Figure 6.11, which shows two sets of histograms, broken out by the levels of each of these two categorical variables—a row for each level.

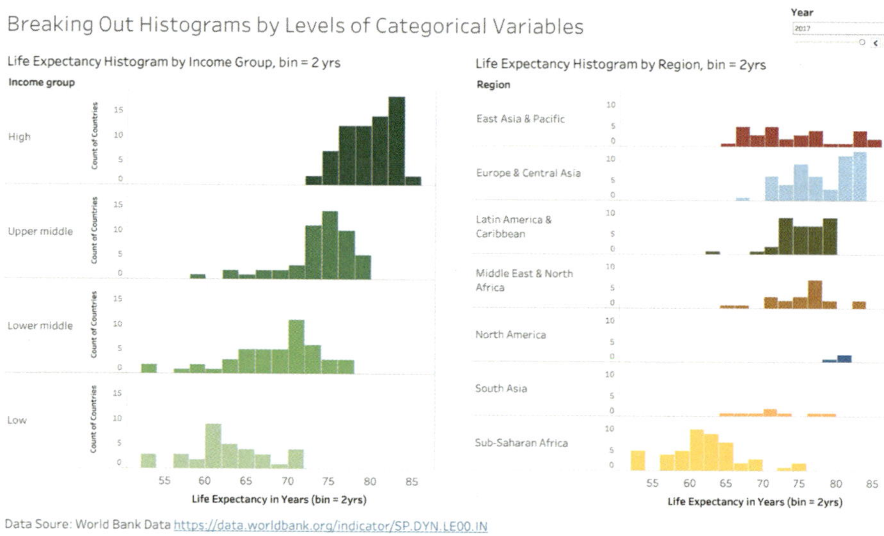

Figure 6.11. Histograms partitioned by the levels of two categorical variables

Notice also that the group of histograms on the left have been broken out by an ordinal variable, income group, and that the rows and color saturations both preserve the inherent order of this variable, with high income at the top and low income at the bottom, as we've seen before.

You've probably heard about the "bell-shaped curve," which is also referred to as the Gaussian or normal distribution. This familiar curve forms a smooth line—a continuous probability function—that's a mathematical function discovered by Carl Friedrich Gauss in 1809 after he and others studied many distributions in astronomy and in games of chance that seemed to form the same shape. You can see that some of our histograms seem to form this familiar shape while others don't.

To show how a histogram relates to the Gaussian distribution, statisticians and chart makers often superimpose a normal curve over the histogram, as shown in Figure 6.12 with our life expectancy histogram. This superimposed curve has the mean and standard deviation of the underlying distribution being graphed.

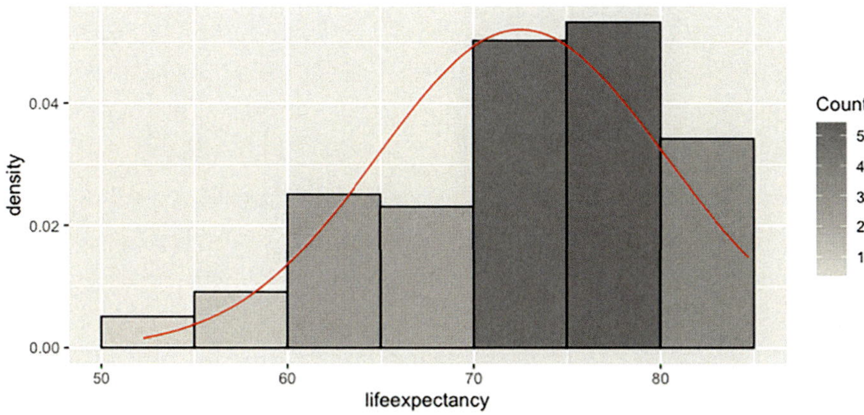

Figure 6.12. A histogram with a superimposed normal curve

Of course very few histograms form a perfectly symmetrical distribution. In the real world, they're often skewed one way or the other. Positively skewed distributions have a longer tail to the right, as shown in Figure 6.13, and this tail "pulls" the mean farther to the right than the median. Negatively skewed distributions are the opposite: they have a longer tail to the left, and this tail pulls the mean farther to the left than the median. The three measures of central tendency are equal to one another for the perfectly symmetric normal distribution in the middle of Figure 6.13.

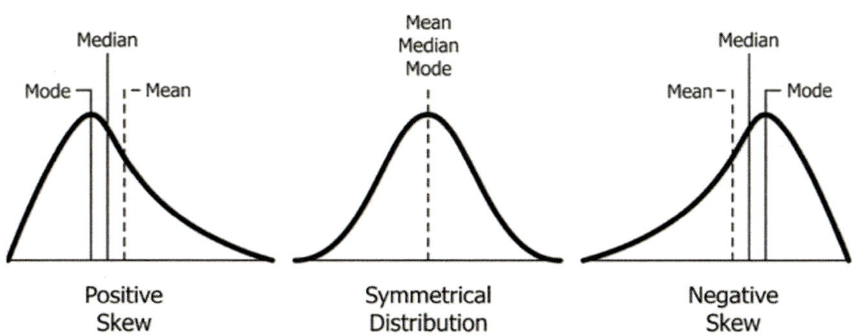

Figure 6.13. Different types of skewness[43]

43. Diva Jain, CC BY-SA 4.0, via Wikimedia Commons

Let's turn our attention to another chart type that conveys both central tendency and dispersion of a quantitative variable: the box plot.

The Box Plot

There's another type of chart that gives us valuable information about the distribution of a set of numbers. It's called the box plot, or sometimes the box-and-whisker plot. While it's very popular in academic research and in the fields of statistics and quality control, I've found that people outside of those fields can sometimes find it confusing or intimidating.

The truth, though, is that it's actually quite simple. As we'll see, it's really just a handy way of summarizing a dot plot by drawing lines and regions, or boxes, that represent its various quartiles. But already some people will be scratching their heads at this point, wondering what a quartile is. So let's start there first, and then the code of the box plot will become clear.

Do you remember when you were at school, and you took a standardized test and got your result back? In the United States, where I grew up, often these nationwide tests told you your percentile. Maybe your score was at the 80th percentile, for example, which meant that you scored as good as or better than 80% of all the other test takers.

Well, quartiles are just special percentiles; percentiles that are multiples of 25%. The first quartile is the 25th percentile—the point in the set that's greater than 25% of the values, and therefore less than 75% of the values. The second quartile is the 50th percentile, and we already know that this value is called the median. And the third quartile is the 75th percentile.

Let's consider a simple, fictional case of thirteen students who took a test on which they could score between 0 and 20 points. They received the following scores: 5, 5, 6, 8, 9, 11, 12, 14, 14, 17, 18, 19, and 20. Figure 6.14 shows where the quartiles would be for these scores. The median, as we know, is the middle value in the ordered set—in this case, a score of 12. That divides the scores into a lower half and an upper half. The first quartile, or 25th percentile, is the median of the lower half, or 8 in this case. The third quartile, or 75th percentile, is the median of the upper half of scores, or 17 in this case.

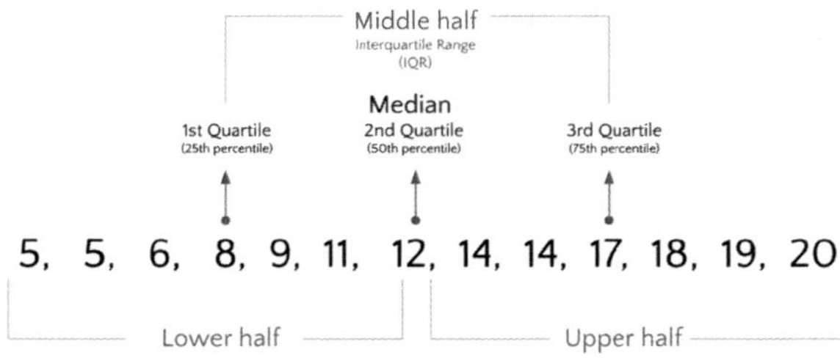

Figure 6.14. Defining quartiles and the Interquartile Range (IQR)

The distance between the first and 3rd quartiles is called the "Interquartile Range," or IQR, and it also contains half of the data. This makes sense, if you think about it, because the gap between 25% and 75% is 50%. So quartiles and the median together help us divide the set of numbers into halves: the lower half, the upper half, and the middle half. Convenient, right?

So how do we apply these concepts of quartiles and the Interquartile Range to real data? How can we combine the statistical measures we've covered with a visual to help us actually see how our data is distributed? The box plot is a very clever invention to do just this. The invention of the box and whisker plot is often credited to John Tukey, an American statistician who wrote about it in 1970. But it's largely based on an earlier form called the Range Bar that Mary Eleanor Spear wrote about earlier in 1952 and again in 1969.[44]

It essentially starts with a dot plot of the values themselves, and it overlays a box for the Interquartile Range that's divided into two at the median. Then it adds straight lines, or "whiskers," outside of the box that extend to the minimum and maximum values in the set. A box plot for our country life expectancy data is shown in Figure 6.15.

44. https://medium.com/nightingale/credit-where-credit-is-due-mary-eleanor-spear-6a7a1951b8e6

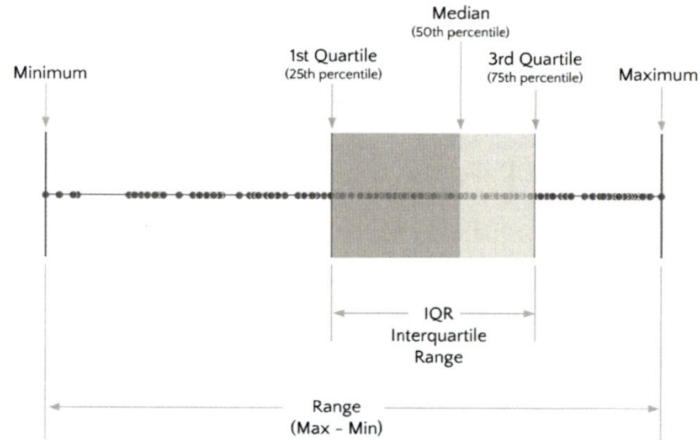

Figure 6.15. The meaning of the lines and regions of the box plot

How does this help us gain a better understanding of the distribution of the data? It divides the data into halves:

- Half of the values are *inside* the box.
- Half of the values are *outside* the box.
- Half of the values are *greater than* the median line inside the box.
- Half of the values are *less than* the median line.

Besides these useful "rules of halves" of the box plot, we can also see that the distribution isn't quite symmetrical; it has a longer whisker on the lower side of the values. We also know that a quarter of the values lie to the left of the left edge of the box. Similarly, a quarter of the values lie to the right of the right edge of the box.

We can put numbers to these figures by adding the continuous horizontal axis for the life expectancy values, as shown in Figure 6.16.

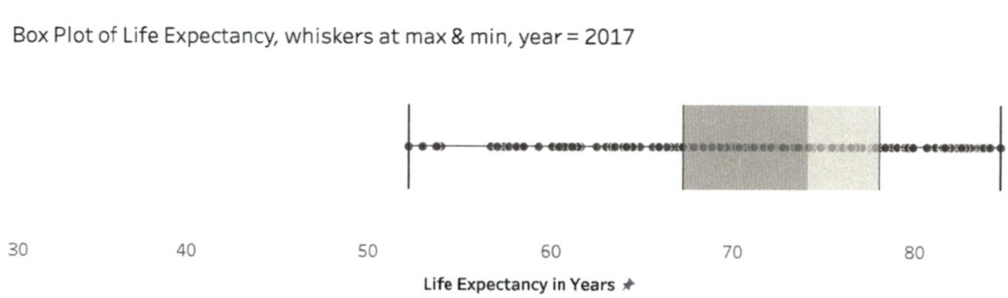

Figure 6.16. The box plot of 2017 life expectancy values by country

There is a lot of occlusion of the dots in the chart. They overlap and block each other to a large degree. This is why many box plots simply omit the dots altogether, as shown in the top version of Figure 6.17. Another approach to deal with this problem is to add "**jitter**"—small and random fluctuations in the vertical position of the dots, as shown in the bottom chart:

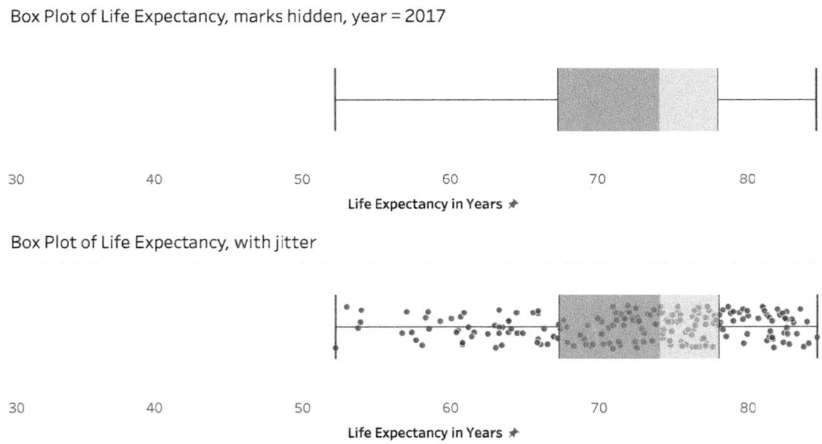

Figure 6.17. Two ways of handling occlusion of the dots

Just like with the histogram, we can add a categorical variable to the view by juxtaposing multiple box plots for an interesting perspective. For example, if we want to know how the distribution of life expectancy differed in 2017 based on country income group, we can break out the dots into four different rows, one for each level of the income group ordinal variable, as shown in Figure 6.18, with the levels in the proper order as before.

There's one important difference in the way some box plots are constructed. In the ones we've considered so far, the whiskers extended all the way to the minimum and maximum values. In this way, it gives a visual perspective of the overall range of the set of numbers.

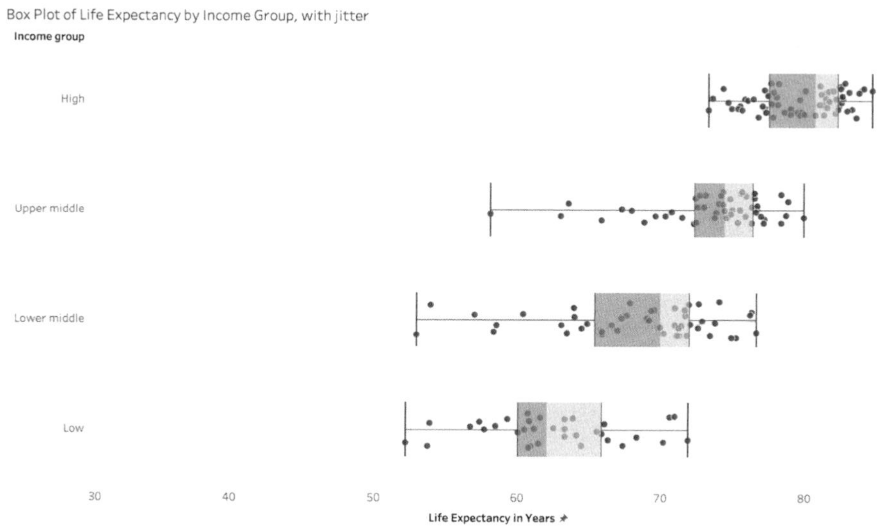

Figure 6.18. Breaking out dot plots by categorical variable

Sometimes, though, we'd like to see whether there are any **outliers** in the data set—values that are significantly different from the others in the set. In these cases, chart makers change the whiskers to extend to 1.5 times the IQR past the edges of the box. If there are no values outside of this distance, then they still stop at the minimum or maximum values.

Take a close look at Figure 6.19 for a version of our life expectancy box plots by income group that have these types of outlier-exposing whiskers:

Note that the high-income and low-income box plots don't look any different than the versions in Figure 6.16. This is because there are no values farther than 1.5 times the IQR past the left and right edges of their boxes. The middle two box plots, though, do look different. Their whiskers on the left side of the boxes stop before reaching the last value. Six country dots are outside of the whiskers. These represent outlier countries, and labels are added to help the reader identify them.

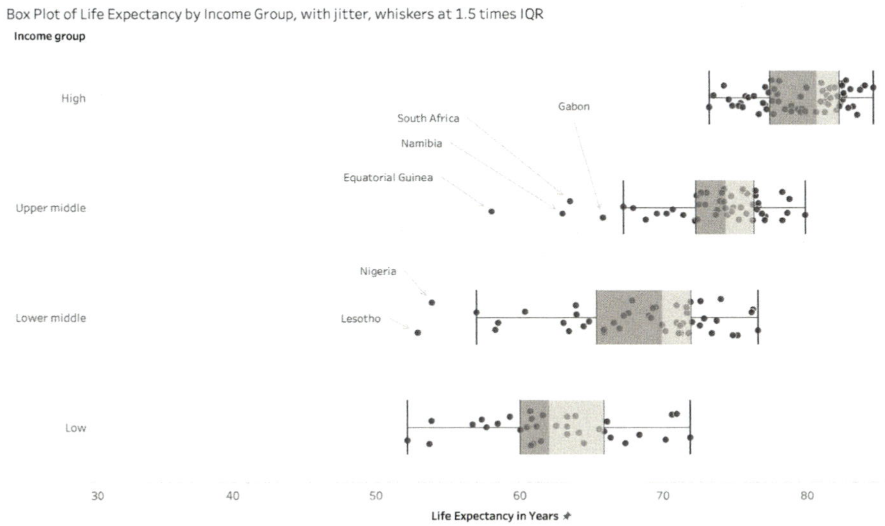

Figure 6.19. A version of the box plot with whiskers extending to 1.5 times IQR

This leads us to yet another Chart Reading Tip:

TIP #12. If there are interesting values in the chart such as outliers, be sure to ask for labels or annotations to help you identify them.

Of course the box plots we have considered so far still have visual channels that the chart maker can use to encode additional variables in the data set. For example, the chart maker can color the dots, adding different saturations for ordered variables or color hues for unordered ones, as shown in Figures 6.20 and 6.21, respectively. Adding color saturation in Figure 6.18 is redundant, since the Income Groups are already separated into rows, so their vertical position already encodes this variable. Encoding the region variable with color hue, though, does add new information to the view: we can clearly see that all of the countries that lie outside of the whiskers are in the Sub-Saharan Africa region.

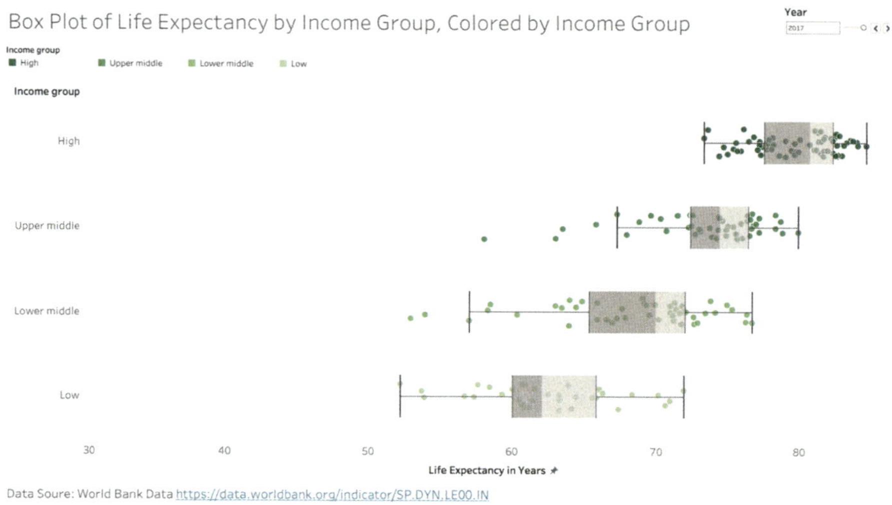

Figure 6.20. Adding color saturation to dots to encode income group

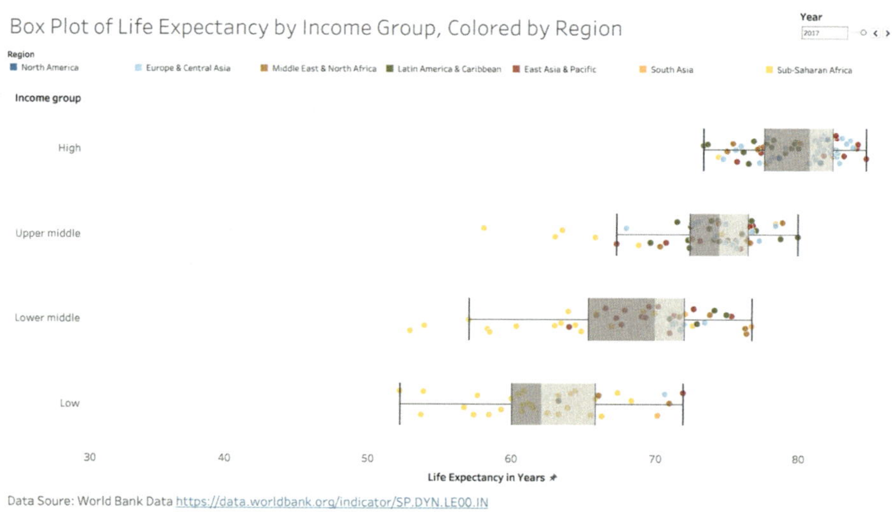

Figure 6.21. Adding color hue to dots to encode region

Lastly, there's nothing stopping the chart maker from using another quantitative variable to encode the area of each of the dots. Sizing dots by population makes it easier to see more populous countries like China, India, and the United States, as shown in Figure 6.22.

Figure 6.22. Encoding dot area with population

Once we spend some time with the box plot, we find that it's not so hard to understand. It gives us a helpful perspective of central tendency and dispersion. It clearly shows the median along with the first and third quartiles, along with the overall spread of the data and any outliers that the set of numbers might contain.

But how does it relate to mean and standard deviation that we talked about earlier in the chapter? An ingenious diagram created by Wikipedia user Chen-Pan Liao connects the box plot and the histogram with the measures of central tendency and dispersion we've discussed. It shows how a perfectly symmetric distribution—the Gaussian or normal distribution—would appear in both visual forms, and how the different measures line up. Notice that since the distribution is symmetric, the mean, median, and mode all line up.

Also notice that the percent of the distribution that lies in the region defined by one standard deviation above and below the mean is 68.27%, which is greater than the portion of readings that lie inside the box of the box plot, which we just learned was half of the values, or 50%.

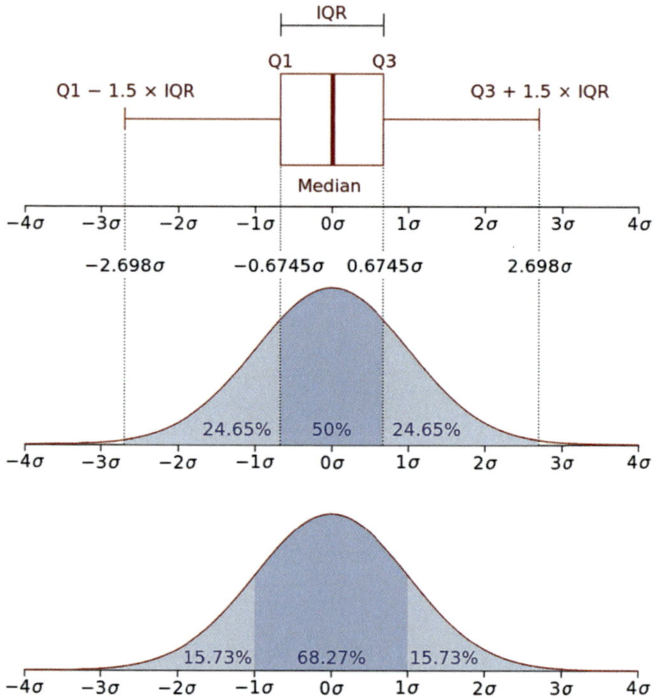

Figure 6.23. A diagram relating the box plot (top) with the normal curve (middle and bottom)[45]

We've been learning about histograms and box plots in this chapter, and how they show the distribution in a set of numbers. I don't know if you noticed, but all of the life expectancy examples and figures we looked at in this chapter were based on data collected in one single year: 2017. So we didn't get any idea how life expectancy has changed over the years.

But there's actually a really inspiring story in this data as it relates to time, and also a sobering reminder in there, too. If we animate the histogram of life expectancy values, playing through the years from 1960 to 2017, the data really comes to life, and we see just how much things have improved for our species on this planet.

We can do the same with the histograms broken out into rows for each region, and see that life expectancies have increased in every part of the world. The box plot can be set in motion as well, and we can watch

45. Jhguch at en.wikipedia, CC BY-SA 2.5 <https://creativecommons.org/licenses/by-sa/2.5>, via Wikimedia Commons

the dots dance across the screen to the right.

We talked about how animations can be eye-catching and inspiring, but that they can be hard to use to answer questions. It's also easy to miss outliers as they disappear almost as quickly as they pop onto the screen. So one approach is to play through an animation to get the audience's attention, and then to leave them with a static display that shows change over time of distribution, as shown in this set of life expectancy box plots in Figure 6.24.

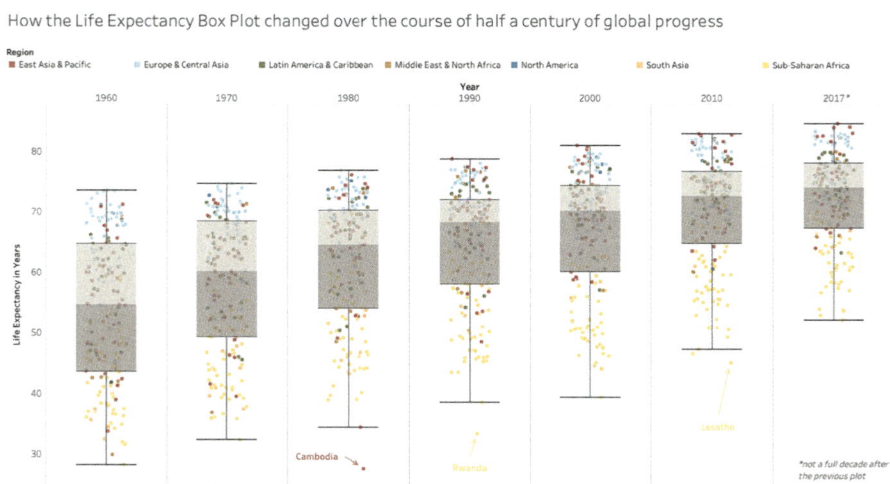

Figure 6.24. A set of box plots that show change of life expectancy distribution over time

Note that the final box plot in the far-right column, which represents data collected in 2017, just seven years after the prior, breaks the decade march from column to column. It's always best that inconsistencies or irregularities such as this are explicitly called out on the chart. In this case, an asterisk next to the column header and a note in the bottom right corner alert the reader.

We can also see clearly by looking at this series of box plots that the median life expectancy rises as the decades progress. In addition, the Interquartile Range gets successively smaller as countries move closer to some natural human lifespan limit, perhaps. The degree of spread in the data has certainly been decreasing.

This version also shows what the eyes of the animation viewer might've missed as the charts progressed quickly through the years—there are some exceptions to the otherwise positive story of progress we've made in the past half century: Cambodia in the 1980s, Rwanda in the 1990s, Lesotho in 2010.

It's a helpful reminder that there are forces in the world that can reverse the incredible progress we're making on the planet. If we can rid ourselves of the type of ethnic and racial strife that results in bloodshed, and if we can root out extreme wealth disparities that create conditions of hunger and disease, we can improve things for everyone, and data can help us tell a story of progress to future generations.

Let that thought play through your mind.

FURTHER LEARNING

Avoiding Data Pitfalls: How to Steer Clear of Common Blunders When Working with Data and Presenting Analysis and Visualizations by Ben Jones (John Wiley & Sons, 2019), ref. Chapter 5—"Pitfall 4: Statistical Slipups": http://avoidingdatapitfalls.com/

Fooled by Randomness: The Hidden Role of Chance in Life and in the Markets by Nassim Nicholas Taleb (Random House Trade Paperbacks; 2nd ed., August 23, 2005): https://www.penguinrandomhouse.com/books/176225/fooled-by-randomness-by-nassim-nicholas-taleb/

"40 years of boxplots" by Hadley Wickham and Lisa Stryjewski (November 29, 2011): https://vita.had.co.nz/papers/boxplots.pdf

"Credit Where Credit is Due: Mary Eleanor Spear" by Ben Jones (August 6, 2019) https://medium.com/nightingale/credit-where-credit-is-due-mary-eleanor-spear-6a7a1951b8e6

Chapter 7

FINDING CORRELATIONS

"Learn how to see. Realize that everything connects to everything else."
–Leonardo da Vinci

In the previous chapter we learned how to see the distribution of a single quantitative variable. Knowing the encodings and chart types that give us a clear picture of the shape of our data is so important because we know that the world contains a great deal of variety in it. And as we saw, the counts, measurements, and readings in our data can vary in interesting ways, too.

In this chapter, we'll learn how to see the nature of the relationship between two or more quantitative variables. The world is also made up of fascinating connections between phenomena. The physical world is an ecosystem that flows and changes together in fascinating ways. Trees in the Amazon pull gallons of water from the soil and pass around 20 billion tons of it into the atmosphere each day, affecting even the level of snowpack in California's Sierra Nevada mountains almost five thousand miles away.

Our interrelated societies contain underlying relationships as well. Information, money, attention, and power change hands continually and produce myriad tendencies and imbalances. Some countries are much wealthier than others, for example, and wealthier countries tend to have lower infant mortality rates. And while we can reason why wealth and health are related in this way, let's leave that question alone for now, and just observe the fact that the two variables are related.

In statistics, when a relationship between two quantitative variables exists, we say they're correlated. We considered **correlation** briefly in the *Data Literacy Fundamentals* book and course in the section on diagnostic data analysis in Chapter 5, but we'll look more closely at it here, especially the visual aspect.

Correlation, as the popular saying goes, doesn't imply causation, but it doesn't rule it out either. It simply conveys the degree to which two variables are associated. We'll consider correlations that are almost certainly **causal** (changes in one variable directly *cause* the other variable to change), those that are **spurious** (they seem causal but they're actually not), and an interesting class of correlations that are based on a third, **confounding** variable that causes both to change.

Often, the most effective way to visualize the correlation between two quantitative variables is to view them at the same time in a scatter plot. Let's dive in and see how this critical chart is made. We'll apply our understanding of the effectiveness of encodings to see why the scatter plot is so effective.

The Scatter Plot

Recall from the first chapter that the most effective way to encode magnitudes is to place marks at positions proportional to their respective magnitudes, and relative to a common baseline. This allows the reader to make the most accurate guesses of their true relative proportions, just like when we use a ruler to measure length. We learned that the dot plot is the chart type that uses the position encoding when showing us the values of a single quantitative variable in our data. And in the previous chapter, we encountered the box plot, which is simply a way of adding quartile reference lines to the dot plot.

If we want to use this same encoding to look at how two variables are related, we can apply the same Cartesian coordinate system we've been considering. We use position along the horizontal, or x-axis, for the first variable, and position along the vertical, or y-axis for the second one, as shown in Figure 7.1.

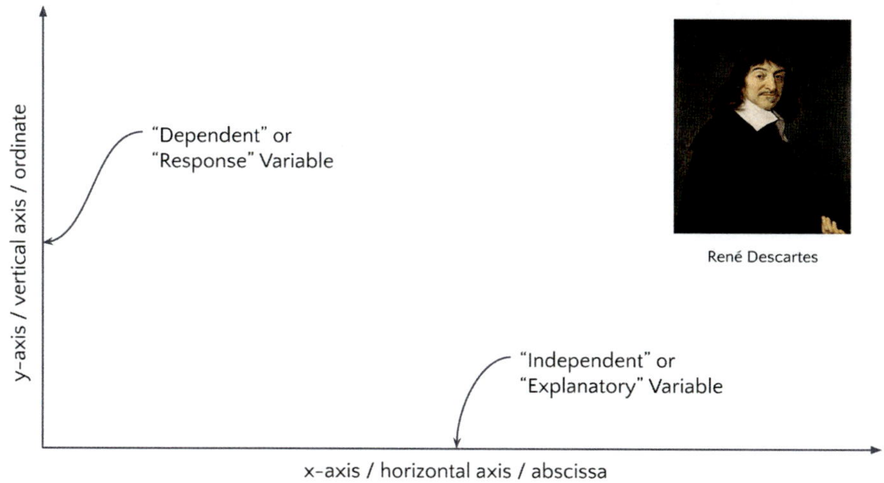

Figure 7.1. The first quadrant of a Cartesian coordinate system

When researchers use this convention, they typically plot the "independent" or "explanatory" variable horizontally, and the "dependent" or "response" variable vertically. This doesn't necessarily mean that changes in the independent variable *cause* changes in the y variable, but rather that in their experiment, they're choosing a value for x as an input and seeing how it impacts a reading taken in y as an output.

For example, if experimenters ask different participants to use their phones for a varying amount of time before going to bed, and then they measure how many hours the participants sleep, then "phone use time" would be the independent variable and "hours of sleep" the dependent variable.

Often we're looking to explore the relationship between variables found in historical data that didn't involve any experiment. We can still use the same convention and plot one variable on each axis.

In the previous chapter, we considered the life expectancy of different countries. What if we were curious about the connection between life expectancy and the degree of urbanization of countries, as measured by the percent of a country's population that lives in a city area? Do countries with a relatively high percentage of city dwellers tend to have longer or shorter lifespans than countries where a high percentage of residents live in suburban or rural areas?

Figure 7.2 is a scatter plot that shows us what the relationship between these two variables used to be in 1960.

Figure 7.2. A scatter plot of percent urban population vs life expectancy in 1960

Each dot represents a country. Dots are positioned horizontally based on the percent of people in each country who live in a city, and vertically by the life expectancy of people born in that country in 1960. For example, Figure 7.3 shows the location of the dot for what's now the Russian Federation (then part of the Soviet Union) to help you understand how its location was determined, and thus how the cloud of dots was formed.

So what does this scatter plot tell us about how urbanization and life expectancy of countries were related in 1960? If we just glance at it, we notice that the dots that are farther to the left look like they're lower down on the grid. The dots on the right are higher up but seem to level off somewhat. It seems like there might be something to this relationship, but we're not quite sure yet. It could just be noise. How do we find out?

Before we learn ways to understand the type and strength of the relationship between two variables, there's one very important caveat to consider. With data, the devil is always in the details, and each variable has its own details.

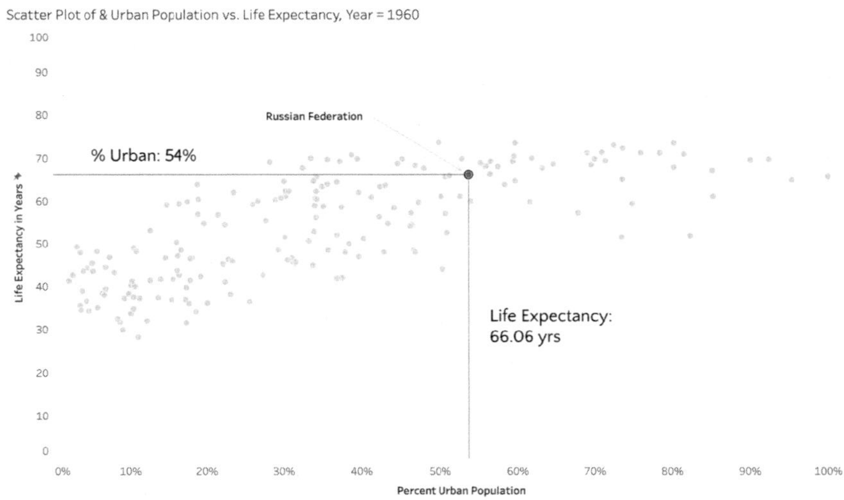

Figure 7.3. The coordinates for the dot that corresponds to the Russian Federation

For example, how do we decide what's an urban area and what's rural? Do we use population density, or presence of certain infrastructure elements like public transportation, or the form of government in a particular place? It turns out that countries use these and other factors to determine their own population breakdown. Here's what we read in the "Details" section of the World Bank website on urban population.[46]

> *"There is no consistent and universally accepted standard for distinguishing urban from rural areas, in part because of the wide variety of situations across countries.... Because the estimates of city and metropolitan area are based on national definitions of what constitutes a city or metropolitan area, cross-country comparisons should be made with caution."*

On top of the inconsistencies in operational definitions of the data, we learn by reading the details section further that the United Nations collects these values from each nation's statistical office each year and then "smooths" them. What exactly is involved with this smoothing, we don't quite know.

46. https://data.worldbank.org/indicator/SP.URB.TOTL.IN.ZS

So we can move forward with our analysis, but we'll have a better appreciation of the fact that we aren't looking at reality itself; we're looking at data collected about reality. There's always a gap between data and reality, a topic I wrote more about in *Avoiding Data Pitfalls*.[47].

This leads us to our next Chart Reading Tip:

TIP #13: Find out the operational definitions of the variables shown in the chart, learn how the records were collected, and consider what potential caveats these details introduce.

Regression

When we look at a scatter plot, we might think we see a certain type of relationship between the x and y variables. But how do we define the type of relationship we see, and how do we know how strong the relationship is? In our example of urbanization and life expectancy, we thought we noticed a general trend that the cloud of dots that was tilted, in a way. How can we take this observation farther?

We can enhance our general observation by adding a trend line to the scatter plot. The trend line, sometimes called the "best-fit line," minimizes the vertical distances from the line to each of the individual dots. There are many different types of trend lines, but a straight line is a common model to apply. This is called **linear regression**, and Figure 7.4 shows the best-fit line as well as a cutout window that gives a "zoomed in" view of this line, along with details about its slope.

The first aspect we notice about the best-fit line is that its slope is positive; that is, the line rises in height as it goes from left to right. Of course the allowable values for urbanization lie between 0% and 100%, so this limits the extent and applicability of the model and its regression line somewhat.

Recall from algebra that the slope of a line is the ratio of the change in vertical position and the change in horizontal position, or, "rise over run." In this case, the line increases in height as it moves to the right.

47. https://dataliteracy.com/avoiding-data-pitfalls/

Since both rise and run are positive, the ratio is positive. And this makes it easy to remember that a line with this kind of slope is indicative of a positive correlation between two variables: higher values of one variable tend to be associated with higher values of the other one.

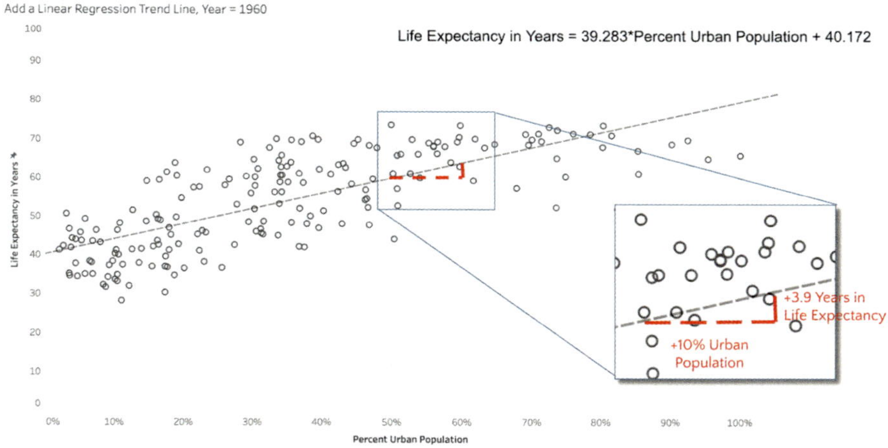

Figure 7.4. A best-fit linear regression trend line added to the scatter plot

This isn't necessarily the case with any two variables, though; some are **negatively correlated** with each other and others have **no correlation** at all. The slope of the best-fit line determines which type of correlation exists, as shown in Figure 7.5. Negative correlations have trend lines with negative slopes, and variables with no correlation have horizontal trend lines with slopes that are close to zero.

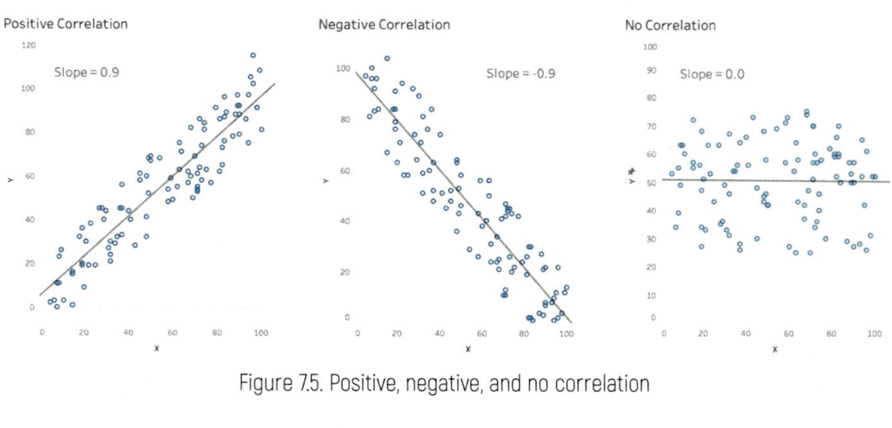

Figure 7.5. Positive, negative, and no correlation

The equation of the best-fit line, Life Expectancy = 39.28 X Percent Urban + 40.17 tells us that, in 1960, countries with an urban population that was 10 percentage points higher than another country tended to have life expectancies that were almost 4 years longer, because 39.28 X 0.10 = 3.928 ≈ 4 years.

For example, the Bahamas was 60% urban in 1960 and had a life expectancy of 64.74, while the United States was 70% urban that same year and had a life expectancy of 69.77. That's a difference of 10 percentage points and 5.03 years in life expectancy, which is pretty close to 4 years more. These models applied to messy, real-world data don't produce exact results.

As the annotations in Figure 7.6 show, though, this comparison of the Bahamas and the U.S. involves countries that lie almost directly on the best-fit line. Not all countries do, though. Some are fairly far away from the best-fit line, actually. And we can find examples of countries with lower urbanization rates but much higher life expectancies than others, such as Norway and Bahrain, as shown.

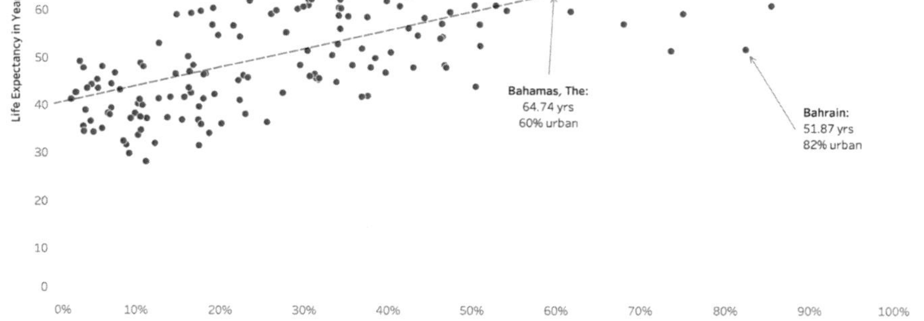

Figure 7.6. Two countries close to the best-fit line and two relatively far away

This illustrates that correlation doesn't necessarily involve universal relationships that always hold for all cases. There can be many other factors at play, and there's also noise in the data. For this reason, it's important to understand how strong or weak a particular correlation is. When two variables have a strong correlation, the individual dots on the scatter plot will lie relatively close to the best-fit line. When they have a weak correlation, they'll be farther away, forming more of a cloud than a discernible line.

This is a very fuzzy criterion, though, so in order to make this assessment more concrete, we calculate the **coefficient of determination**, or R^2 (sometimes "R^2" or "R-Squared"). A strong correlation has an R^2 value that approaches 1 while a weak correlation has a R^2 that's closer to 0. Two variables that are perfectly correlated have an R^2 value of 1.0, meaning that changes in the value of the x variable determine 100% of the variation in the y variable. An R^2 of 0.50 means that 50% of the variation in the y variable can be accounted for by changes in the x variable. Figure 7.7 shows what correlations of different strengths look like.

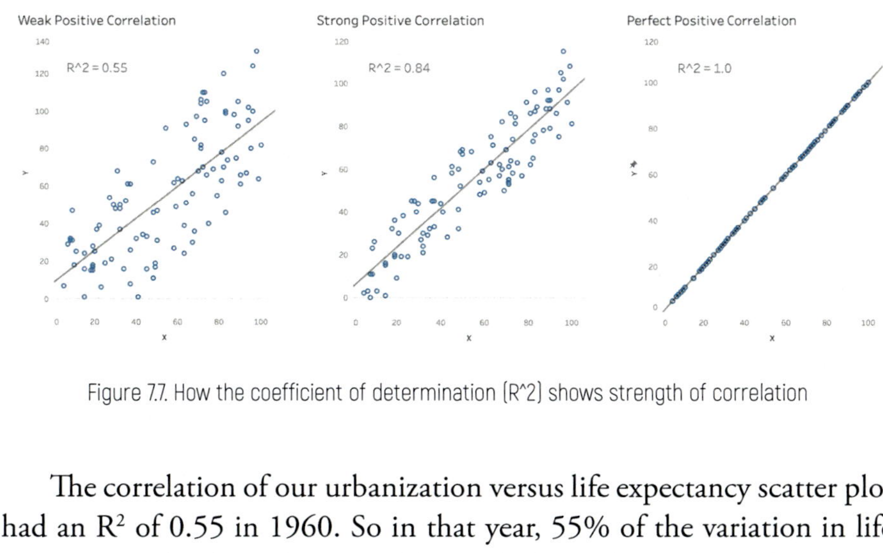

Figure 7.7. How the coefficient of determination (R^2) shows strength of correlation

The correlation of our urbanization versus life expectancy scatter plot had an R^2 of 0.55 in 1960. So in that year, 55% of the variation in life expectancy (the dependent variable) could be explained by differences in urbanization (the independent variable). This isn't a particularly strong relationship. In the social sciences, it's common to consider R^2 values between 0.25 and 0.64 as moderate. So we'll say there was a moderately strong positive correlation between these two variables.

It's important at this point, though, to remind ourselves that a moderate or even strong correlation doesn't prove that one variable *causes* the other. A country can't say, for example, that if they move their citizens who live in rural areas to the cities, that this will suddenly cause their life expectancy to rise. Correlation just tells us that there's a relationship. It doesn't say that the relationship is necessarily causal.

Some correlations are obviously causal. Your gas bill is probably lower in the summer months because the temperature outside is higher, and you don't run your heater in those months. This is a negative correlation (as one goes up the other goes down), and there's a good reason to argue that one causes the other.

But we can find examples in which two variables are definitely correlated, but instead of one causing the other, they're both caused by a third, "lurking variable," often called a **confounder**. For example, months of the year that have high numbers of home renovation permit applications in Seattle also tend to have a high volume of calls to complain about rat sightings in New York City. Neither causes the other, in this case; that would be ridiculous.

Rather, they're likely both caused by a third variable: weather. People are less likely to want to start home renovation projects in the cold, rainy winter months of Seattle, and rats (and people!) in New York are more likely to be huddled away from the snow and ice in those same months, out of sight and out of mind. Correlation does not imply causation.

Before we end this section on regression, it's worth considering an example of a highly nonlinear relationship between two variables. Linear regression isn't the only type of regression. Two variables can be correlated in many different ways. As a case in point, let's consider the relationship between a country's standard of living and its rate of infant mortality (deaths per 1,000 live births). If we create a standard scatter plot of GDP per capita versus infant mortality rate in 2018, we get the scatter plot on the left of Figure 7.8.

Notice the deep curve in the shape of the data in the scatter plot on the left? It isn't even close to a straight line. Small increases in GDP per capita on the left side of the chart result in huge drops in infant mortality. Large increases in GDP per capita on the right side of the chart result in very small reductions in infant mortality. The trend line is now a best-fit *curve* rather than a best-fit line.

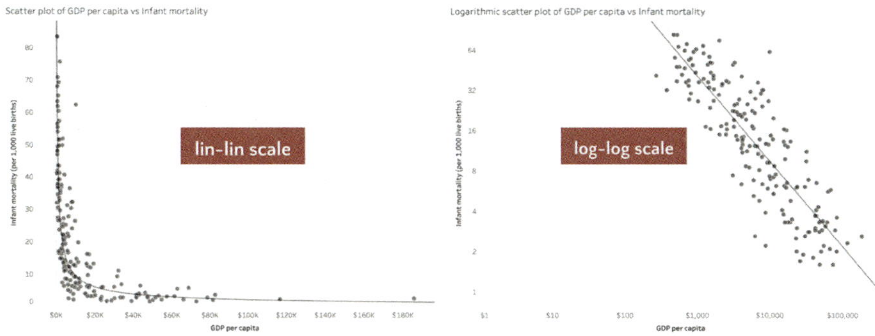

(a) (b)

Figure 7.8. Linear axis scales vs. logarithmic axis scales

These two variables follow a **power law** relationship instead of a linear one. What does this mean? It means that one of the variables varies with the other raised to an exponent, or power. If we change both of the axes from linear scale to logarithmic scale, then the exact same regression model can be drawn with a straight line, as shown on the right of Figure 7.8.

Logarithmic scales can be confusing and disorienting for even many seasoned chart readers. That's because the way the axis is formed is fundamentally different from the standard linear scale. On the left of Figure 7.8, notice how the linear x-axis goes from \$0 to \$20K to \$40K, and so on. Each tick mark increases by the same amount: \$20K. But with the logarithmic scale on the right of Figure 7.8, the x-axis doesn't behave like this at all. It goes from \$1 to \$10 to \$100 to \$1,000 and so on. Each equally spaced tick mark increases by a *factor of ten*. The same transformation was applied to the y-axis in the version on the right, except that vertical tick marks increase by a factor of 2 (1, 2, 4, 8, etc.) instead of a factor of 10.

But what does all of this mean for the actual relationship between these two variables in 2018? Well, if we look even more closely at the log-log scale version of the scatter plot in Figure 7.9, we see that a *10-fold increase* in GDP per capita seems to come along with a roughly *4-fold decrease* in infant mortality. We can see that by looking closely at the graph, and we can confirm it by considering the equation for the power law model: $y = 3437x^{-0.64}$. If we increase x by a factor of 10, y changes by a factor of $10^{-0.64}$, which equals 0.25, or ¼: a 4-fold decrease.

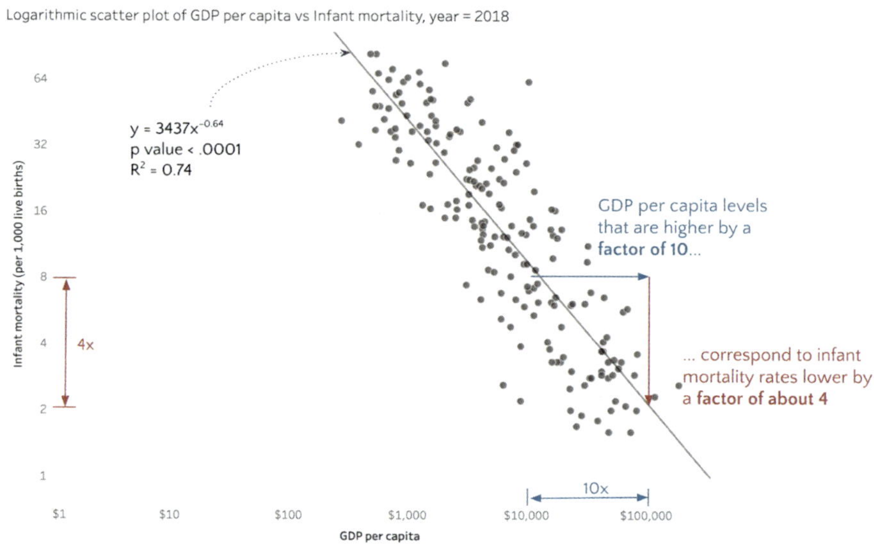

Figure 7.9. A closer look at the log-log scale scatter plot

There are many other types of relationships we can find when we compare two quantitative variables against each other. Linear and power law relationships are just two of them.

This leads us to our next Chart Reading Tip:

TIP #14: If a model like a regression line has been applied to the data in the chart, determine what type of model it is, and how well it fits the data.

So far, our scatter plots have used two channels to encode our data in visual form: horizontal position and vertical position. As you know by now, there are other visual channels that we can use to enrich the display to include even more of our data. Let's consider how we can take the scatter plot to the next level.

The Bubble Chart

The scatter plots we've considered are helpful in that they show us the relationship between two quantitative variables in our data. And this chart type is brilliantly constructed using the single most effective encoding channel, position on a common scale, not once but twice! They take advantage of the fact that a page or screen is a two-dimensional surface, and they use horizontal position to encode one variable, and vertical position to encode the other.

But so far the scatter plots all included circles of the exact same size. Here's an opportunity to add a third variable to view by using it to encode the area of each circle. Recall, though, that area is lower down on Munzner's effectiveness list than position, so we'd only be following the effectiveness principle if we used a variable that's less important to our consideration for this lower channel. In a scatter plot, we want to see the two most important variables encoded with the most effective channel: position.

So which variable would be appropriate to use to encode the size of each country's circle? Technically we can use any ordered variable we want: ordinal, interval, or ratio scale. But the expressiveness principle suggests that it would be a good idea if it were somehow related to the size of the country itself, such as land area or population. Let's use population, and the scatter plot turns to a bubble chart, as shown in Figure 7.10.

Figure 7.10. A bubble chart of urbanization and life expectancy, sized by population

We can see larger circles that we know are associated with the more populous countries like China, India, and the United States, and we see smaller circles that represent countries with fewer people. The circles are semitransparent now since some of the larger circles occlude some of the smaller ones, and they have white borders so that the edges of overlapping circles are easier to see.

But we're not done yet! All of these circles are the same gray hue. We've used color in two ways before, and we can do so again. We can use color saturation to encode a fourth quantitative variable in the display. In Figure 7.11, we see circles that are a higher saturation of blue have a higher *overall* urban population (measured in number of people, not percentage).

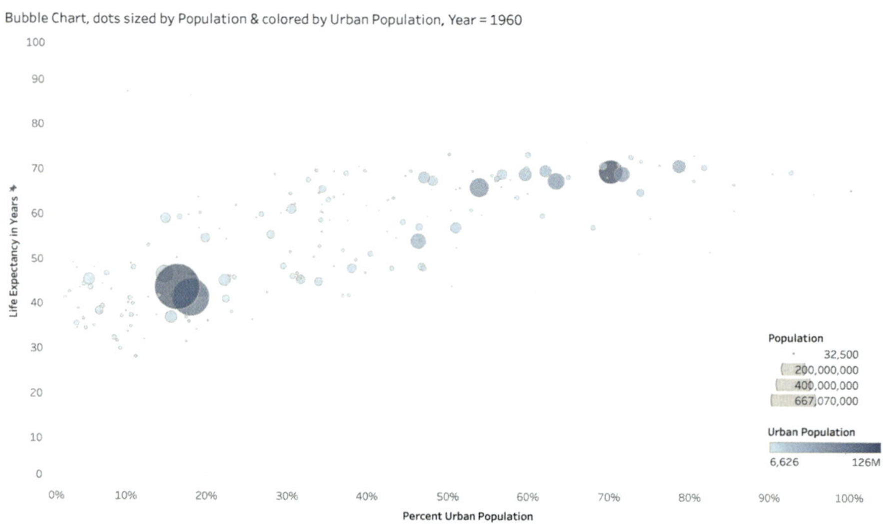

Figure 7.11. A bubble chart with color saturation encoding a fourth quantitative variable

If you think about it, this fourth quantitative variable, overall urban population, can actually be mathematically derived from two of the other variables in the chart: population (circle area) times percent urban population (horizontal position).

It increases our depth of understanding, though, because we see that even though China and India, the two largest circles to the left on the chart, are relatively low in the percent of their populations that live in a city, in absolute terms they have way more city dwellers than many of the less populous countries in the view. This version of the chart dispels the incorrect notion we may have had that all of the countries to the left in the chart have hardly any city dwellers.

Of course, as we've seen before, we can use color in a different way. Color saturation is expressive of ordered variables like urban population, and color hue is expressive of unordered variables like region. If we color the circles based on their region, we get the version of the bubble chart shown in Figure 7.12.

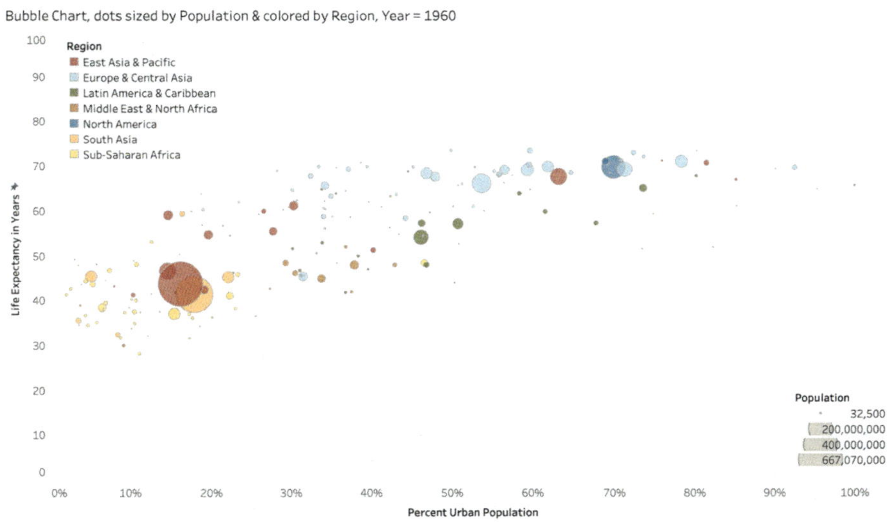

Figure 7.12. A bubble chart with color hue encoding a categorical variable, region

This version of the bubble chart has two legends: a color legend in the top left corner and a circle area legend in the bottom right corner.

This leads us to yet another Chart Reading Tip:

TIP #15: Pay close attention to the different legends in the chart and see how they reveal encodings other than position such as size, color, or shape.

All of the scatter plots and bubble charts we have considered so far in this chapter have shown us the relationship between two variables in one period of time, namely, a particular year:1960. But what did the relationship between urbanization and life expectancy look like more recently, and did the relationship change over time?

Correlations Over Time

We can use some of the same techniques we've used so far in the book to see how correlation has changed over time. For example, we can place two bubble charts side by side, one for 1960 and the other for 2017. Figure 7.13 shows us this exact arrangement, with the 2016 chart on the left, consistent with the left-to-right convention we learned about in Chapter 5.

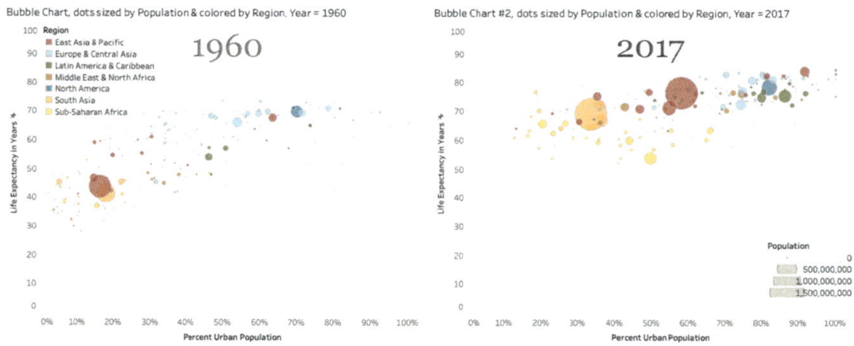

Figure 7.13. Two juxtaposed bubble charts showing different years of data

Arranging the two charts horizontally like this allows us to see that, in general, the cloud of circles rose vertically to higher life expectancy levels over this time. Since the two charts share a common horizontal baseline, and since the y-axes have the same maximum value of 100 years, we can compare heights of circles across the charts. This choice of alignment makes it more challenging to compare horizontal positions of the circles in the two different charts, though, because they don't share a common vertical baseline of 0% urban population.

This choice has other drawbacks, as well. These juxtaposed charts say nothing about what happened in the years *between* 1960 and 2017. On top of that, they make it difficult to tell how any one country changed from one year to another. For these types of comparisons, we can decide to instead view a connected scatter plot, as shown in Figure 7.14. This chart shows how the relationship between urbanization and life expectancy for the ten most populous countries changed year by year.

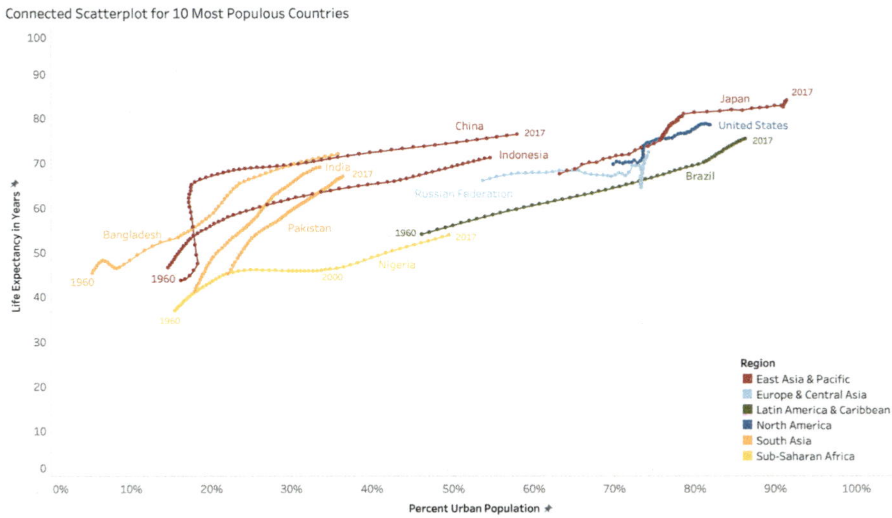

Figure 7.14. A connected scatter plot of urbanization and life expectancy for select countries

This interesting chart type incorporates change over time, but not on either the horizontal or vertical axis. The two position encodings are reserved for our two quantitative variables, as they were in the basic scatter plot. Instead, each dot represents a point in time, and the series of dots for each country moves in the two-dimensional space defined by their respective urbanization rates and life expectancy levels. The lines connect dots for subsequent years: the 1960 dot is connected to the 1961 dot, which is then connected to the 1962 dot, and so on.

This chart construction can sometimes be disorienting for many, as left-to-right doesn't necessarily correspond to the progression of time. In the example in Figure 7.14, the dots do generally move from left to right and from bottom to top, but they could easily turn around and move in the opposite direction.

Consider, for example, the wrinkle in the light blue line for the Russian Federation, where life expectancy rates evidently dropped for a few years starting in 1991, when the Soviet Union dissolved. Connected scatter plots can work really well, but they can sometimes be confusing or even unreadable, depending on the path the dots take over time.

We might instead ask to be shown an animated version of the bubble chart, with the bubbles moving as time "plays" from the past to the present. This approach was made famous by Hans Rosling, the Swedish physician, academic, and public speaker whose TED talk titled "The best stats you've ever seen"[48] currently has over 14 million views, and is a must-watch for anyone interested in the art of communicating data to an audience.

You can watch a video of the urbanization versus life expectancy bubble chart on YouTube[49], or you can see the superimposed versions of the yearly positions in Figure 7.15.

Figure 7.15. A chart showing each circle's yearly positions of the animated bubble chart

The final way we'll consider to view how two different quantitative variables changed over time breaks away from the scatter plot convention of placing the two variables on perpendicular axes, and returns us to more familiar territory: the line chart. We place time going from left to right on the horizontal axis, and we superimpose both quantitative variables on the same chart, each with their own vertical axis. Both can be lines, or, as shown in Figure 7.16, one can be a line and the other a set of bars.

48. https://www.ted.com/talks/hans_rosling_the_best_stats_you_ve_ever_seen
49. https://youtu.be/mbN8fBOKHWk

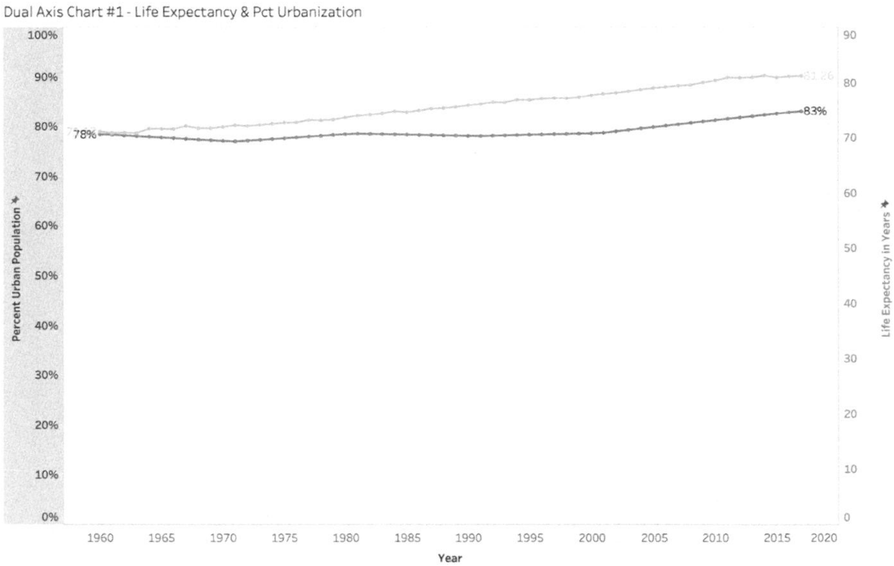

Figure 7.16. A dual axis chart showing a superimposed line chart and bar chart

This chart form is known as a **dual axis chart**, and it shows us how both life expectancy and urbanization changed over time in the United Kingdom. As clear and effective as it might seem to be, it can actually be quite misleading, and many data visualization experts advise against using it altogether.[50]

Why is that? There are two main drawbacks with this type of chart. The first is obvious: it's really only possible to look at these two quantitative variables for one series at a time: in this case the UK. Can you imagine trying to compare multiple lines with multiple bars? That would be chaos.

The second flaw is less obvious. Looking at this first version of the dual axis chart, it seems that life expectancy grew faster than urbanization in the UK since 1960. It seems this way because the lines start together on the left, and then the blue line for life expectancy rises faster than the gray line for urbanization. Even expert chart readers will jump to this conclusion after seeing this first version.

Compare it, though, with another version of the dual axis chart shown in Figure 7.17. This second version charts the exact same data,

50. https://blog.datawrapper.de/dualaxis/

but it seems to make the opposite case. These two quantities look like they were both growing relatively slowly, and then urbanization surged ahead starting in the early 2000s.

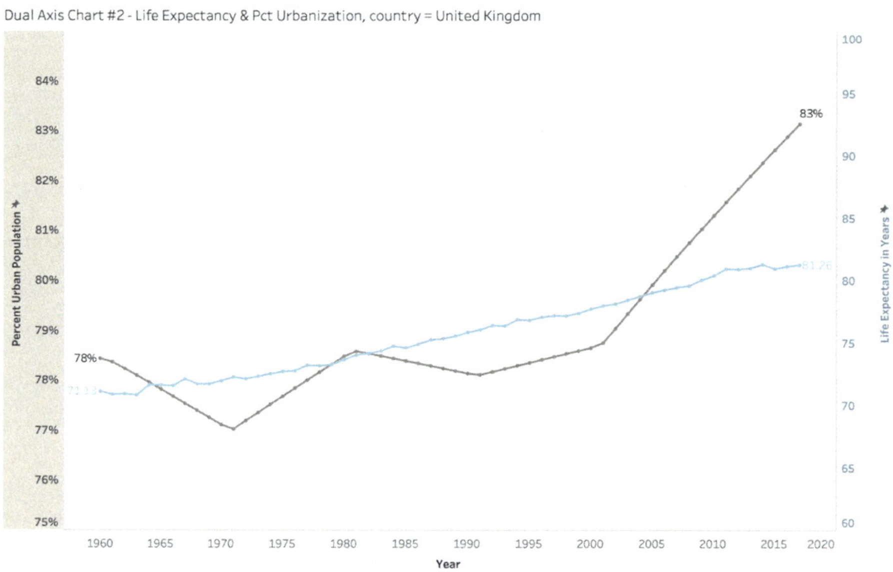

Figure 7.17. A second version of the same dual axis chart with different axes limits

The only differences between these two versions of this chart are the starting and ending points of the two vertical axes. Take a look at both versions and compare the tick mark labels on each of the axes. It's these choices, not the data itself, that will determine where the lines cross, and how steep their slopes appear relative to one another. For this reason, we need to be careful with this chart type, and make sure that we don't form our understanding based on arbitrary configuration choices instead of the data itself.

FURTHER LEARNING

Video: "The best stats you've ever seen" by the late, great Hans Rosling: https://www.ted.com/talks/hans_rosling_the_best_stats_you_ve_ever_seen

Presentation: "The Art of Honest Theft" by Hannah Fairfield, Tapestry Conferecen 2013: https://www.ted.com/talks/hans_rosling_the_best_stats_you_ve_ever_seen

Blog Post: "Why not to use two axes, and what to use instead" by Lisa Charlotte Rost, on the Datawrapper blog, May 8, 2018: https://blog.datawrapper.de/dualaxis/

Paper: "The Connected Scatterplot for Presenting Paired Time Series" by Haroz, Kosara & Franconeri (2016): https://eagereyes.org/papers/the-connected-scatterplot-for-presenting-paired-time-series

Chapter 8

READING AND INTERPRETING MAPS

"A map, it is said, organizes wonder."
–Ellen Meloy

If you've ever looked up at a starry sky on a clear, moonless night, you've had that feeling of wonder about the immensity of the universe, and of our small place in it. As the epigraph of this chapter states, this sense of wonder at our physical surroundings is infused in the maps we create to represent those surroundings for both useful and artistic purposes.

Figure 8.1. Ben Alberts at Sand Point, Washington, July 2020. Photo by Ben Jones

More than just inspirational for ancient humans, the stars were also practical and necessary for essential tasks like navigating and telling time. Astronomy is said to be the oldest of the natural sciences. Some of the earliest known charts are maps of the stars in the heavens. In the southwest of France, the caves of Lascaux contain dots of constellations including the Pleiades cluster dating back to around 14,500 BCE. Cave drawings in Spain dating from around 12,000 BCE show the Corona Borealis constellation.

The oldest known maps of the Earth are from ancient times, too. Ancient humans etched paths, mountains, and rivers on cave walls and mammoth tusks that time has preserved over the ages. Some of these ancient maps are thought to help prepare for a hunt, or tell the story of a particularly successful one. One seems to show the plan of a Neolithic village. These maps all show us that the ability of our species to capture and convey geographic information in encoded, graphical form is far older than civilization itself.

A Primer on Maps

Fast forward to our day, and we find ourselves in a world where we use maps all the time. Whether we're finding the best route to get home from work, checking the weather report before a planned vacation, or understanding how our country voted in an election, maps are a common tool for most of us. Technology has enabled our addiction to maps: we can hold in our pocket the satellite images of 98% of the populated surface of the Earth.[51]

Between ancient times and our day, there's a vast and impressive record of the development of maps for various purposes. The history of cartography can't be told without mentioning the Babylonians and Egyptians, the Greeks and Romans, the Chinese, the Arabs and Persians, the Polynesians, and Europeans in medieval times as well as during the Renaissance, to name a handful of the pioneering civilizations of the field. Our goal, however, isn't to tell the history of maps, but to better understand how to use them.

One cartographer of note, though, is Gerardus Mercator, a 16th-century Flemish cartographer. His was an era of nautical exploration known as the

51. https://www.blog.google/products/maps/google-maps-101-how-imagery-powers-our-map/

Age of Discovery. And Mercator wasn't just a cartographer; he was also a businessman. He made an income selling globes and maps around the world, and he eyed an opportunity to address an unmet need. He developed and published what we now call the Mercator 1569 world map, as shown in Figure 8.2.

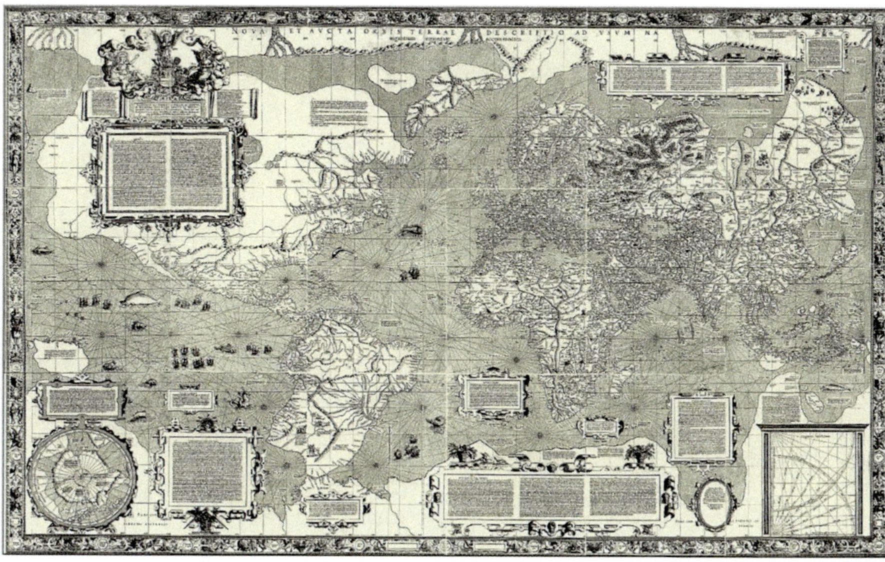

Figure 8.2. Gerardus Mercator's 1569 world map[52]

According to the English translation of the Latin title of his map, Mercator claimed to have created a "new and more complete representation of the terrestrial globe properly adapted for use in navigation." His goal was to create a map that would be useful for sailors as they made their way across the ocean. The innovative feature of the map was that any ship traveling with a constant bearing (that is, along a rhumb line) could track their progress using a straight line on his map.

His map was slightly ahead of its time, though, and didn't catch on until the middle of the nineteenth century, when other inventions made it easier for sailors to determine their position at sea. Its subsequent adoption and popularity, though, was met with an intense dislike on the part

52. Wikimedia Commons, Public Domain: https://commons.wikimedia.org/wiki/File:Mercator_1569.png

of professional cartographers. Experts both then and now point out that in order to preserve *angles* in the way that it does, Mercator's map has to exaggerate the *areas* of land masses that are far away from the equator. And showing the polar regions in a useful way just isn't possible.

So its popularity waned in the 1900s due to these criticisms, but it surged again in more recent years due in part to the ease of alignment of satellite image tiles. For better or worse, it's cousin the Web Mercator is currently the de facto standard for web mapping applications like Google Maps and Microsoft's Bing Maps.[53]

This story highlights the challenge inherent to the task of depicting a globe on a flat surface. A globe is a three-dimensional object, and most maps are shown on two-dimensional surfaces. How do we show a three-dimensional object on a two-dimensional surface? We do so imperfectly, and with a map projection. A **map projection** is a flattening of a globe's surface onto a plane, like an orange peel onto a table. Like the many shapes the flattened orange peel might take, there are many different map projections that cartographers have created over the years.[54] Mercator's map is just one of many in existence today.

To briefly get a taste for some of the more common ones, let's start by imagining the globe covered with circles that run parallel to the equator, called lines of **latitude**, and arcs that connect the North and South Poles, called meridians, or lines of **longitude**. Figure 8.3 shows what this looks like.

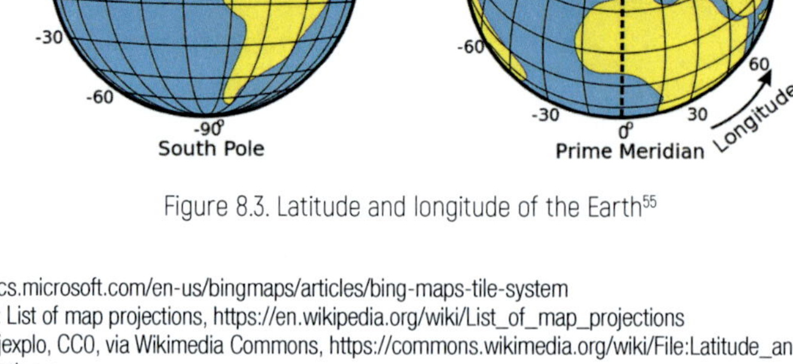

Figure 8.3. Latitude and longitude of the Earth[55]

53. https://docs.microsoft.com/en-us/bingmaps/articles/bing-maps-tile-system
54. Wikipedia: List of map projections, https://en.wikipedia.org/wiki/List_of_map_projections
55. Source: Djexplo, CC0, via Wikimedia Commons, https://commons.wikimedia.org/wiki/File:Latitude_and_Longitude_of_the_Earth.svg

Now, we can compare six different map projections, as shown in Figure 8.4. The parallels and meridians form a network of **graticules** that show how each map projection represents the globe. None of these projections are perfect. Each one distorts either angles or areas, or some combination of the two.

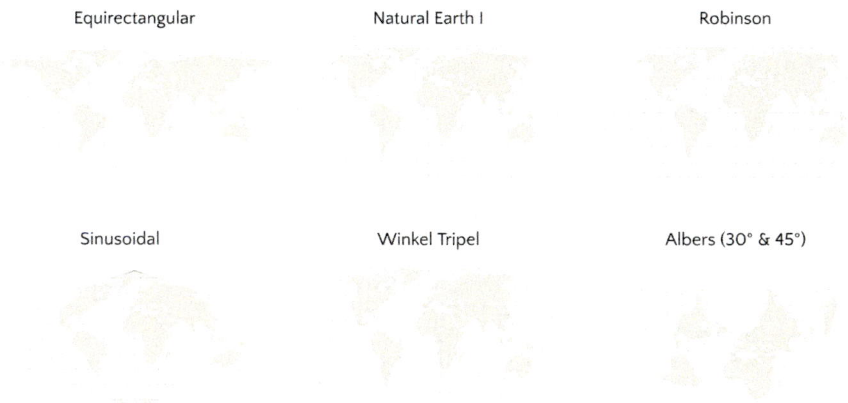

Figure 8.4. Six different map projections

What's powerful about them all, though, is that they create a canvas on which we can paint the geospatial variables in our data. Doing so can reveal to us the all-important answer to the question "where?" In the following sections of this chapter, we'll consider how to enhance maps with each of our common mark types: 0D points, 1D lines, and 2D shapes.

The Symbol Map

We already know that position is the most effective encoding channel for magnitudes. The dot plot and the scatter plot both make use of the position encoding to enable us to make accurate comparisons between values. Maps, by their very nature, allow us to use position to encode a mark's location relative to certain baselines.

With maps of the Earth's surface, as we've been considering, there are two baselines that we use to specify an object's location on the surface of the planet, ignoring elevation for the time being. These lines are both shown in Figure 8.3. One is the equator, an imaginary line drawn around the Earth defined by points that are the same distance from both poles. The other is the Prime Meridian, an arc that connects the North Pole to the South Pole and runs through Greenwich, England, where the Royal Observatory is housed.

From these two reference lines, we can specify the location of any place on the surface of the Earth using geographic coordinates that have two components: latitude and longitude.

- **Latitude** gives the north-south position as an angle that ranges from 0 at the equator to 90 (or, 90°N) at the North Pole and –90 (or, 90°S) at the South Pole.
- **Longitude** gives the east-west position relative to the Prime Meridian, which has been (arbitrarily) assigned a longitude of 0°. The longitude of other places is measured as the angle east or west from the Prime Meridian, ranging from 180 (or 180°E) to the east and –180 (or, 180°W) to the west.

What does this look like on a flat map, though? Figure 8.5 shows the coordinates of the center points of five different countries on the Web Mercator map. The center point of one of the countries shown, the United Kingdom, lies very close to the Prime Meridian: it has a longitude of –1.3. The center point of another country shown, Gabon, lies almost perfectly on the equator, and has an latitude of –0.41. The center point of the United States has coordinates of 40.08, –98.98 (or 40° 4' 48"N, 98° 58' 48" W in DMS [degrees minutes seconds] format). The red lines with arrows show how each of these coordinates relate to their relative reference line.

If you think about it, these lines create something very similar to the Cartesian coordinate system used by the scatter plot and bubble chart we learned about in the previous chapter. The equator is an x-axis of sorts, and each point's horizontal position on the grid is the east-west location defined by its longitude. The Prime Meridian is a y-axis of sorts, and each point's vertical position is the north-south location defined by its latitude.

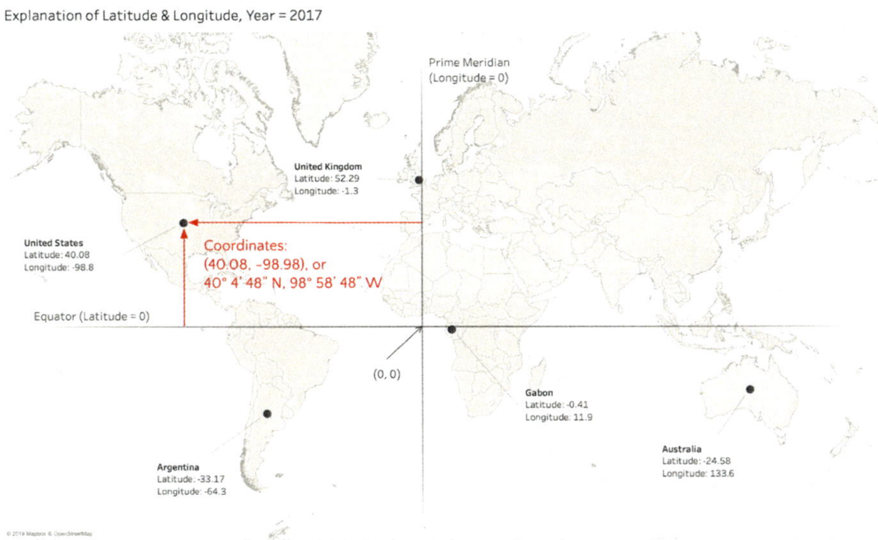

Figure 8.5. A Web Mercator map showing the coordinates of five country center points

So, just like the scatter plot and the bubble chart place shapes within a Cartesian plane, the **symbol map** places shapes on the two-dimensional map projection. This applies at the local level as well as the global level. The chart on the left of Figure 8.6 looks like a typical bubble chart, with circles of various sizes positioned on a plane. When we reveal the satellite image of the associated terrain on the right of the figure, though, we see that the circles are actually locations of building sites in downtown Seattle.

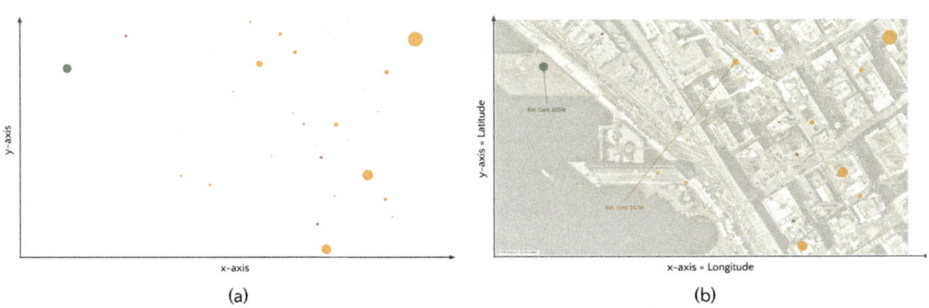

Figure 8.6. The similarity between the bubble chart (a) and the symbol map (b)

Like bubble charts, symbol maps can encode more than just position. Once the shapes have been placed on the map, the size and color channels are still available to encode with other variables in our data. For example, the symbol map showing country population by region in Figure 8.7 uses all four of these encoding channels.

Web Mercator Symbol Map of Country Population, colored by Region, 2019

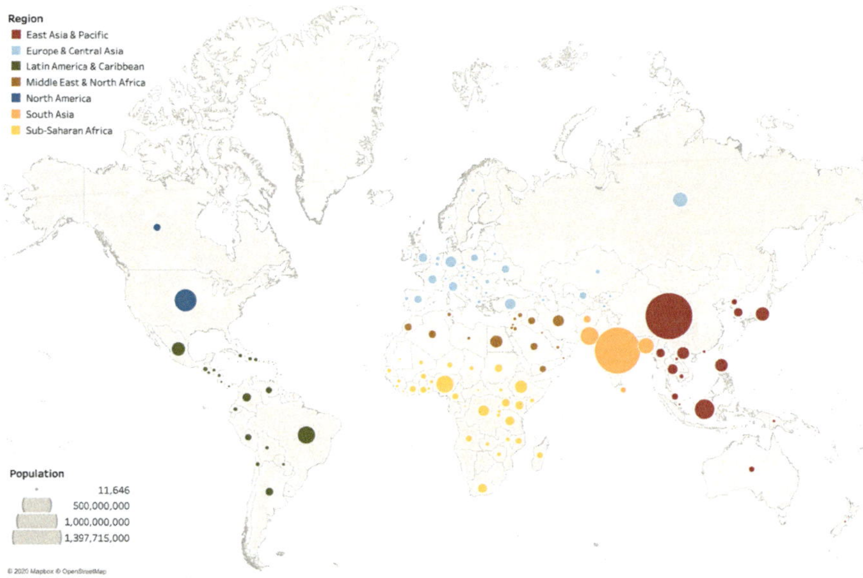

Figure 8.7. Symbol map of country population shown on the Web Mercator map projection

To summarize, this map encodes four variables with four different visual channels:

1. Horizontal position = Longitude of country center point (quantitative)
2. Vertical position = Latitude of country center point (quantitative)
3. Circle area is proportional to population (quantitative)
4. Circle color hue is mapped to region (categorical)

Maps whose symbols vary in size proportional to a quantitative variable are sometimes called **proportional symbol maps**. Compare this first version of the symbol map, which uses the Web Mercator map projection, with the six different versions shown Figure 8.8, which show the exact same data on six different map projections.

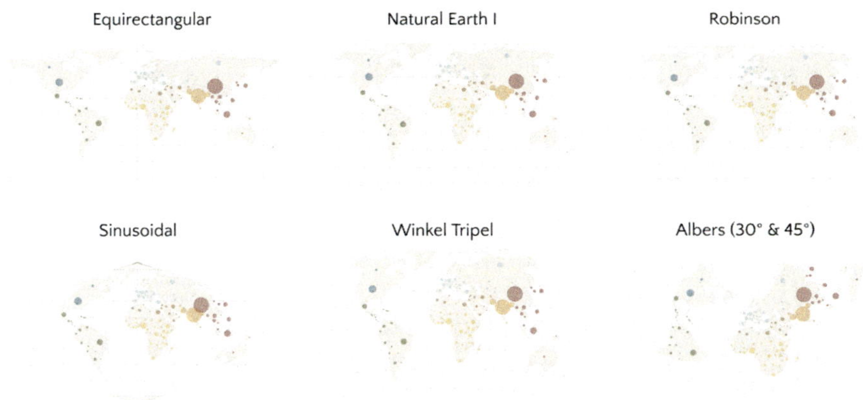

Figure 8.8. Country population symbol map in six different map projections

In particular, pay attention to the relative areas of the countries in the different projections, especially those to the north, at the tops of the maps. In the Web Mercator version in Figure 8.7, Greenland seems almost as large as the entire African continent. In reality, the surface area of the African continent (11.73 million square miles) is over 14 times greater than that of Greenland (836,300 square miles). This is the landmass area distortion to which critics of the Mercator often refer. For this reason, many experts don't like showing global data on the Mercator projection. The other projections in Figure 8.8 don't have anywhere near this same level of area distortion.

Let's switch to the natural Earth projection, and, instead of giving circles different color hues for each Region (a nominal variable), let's give them different color saturations for Income Group, an ordinal variable. Figure 8.9 shows our updated country population map.

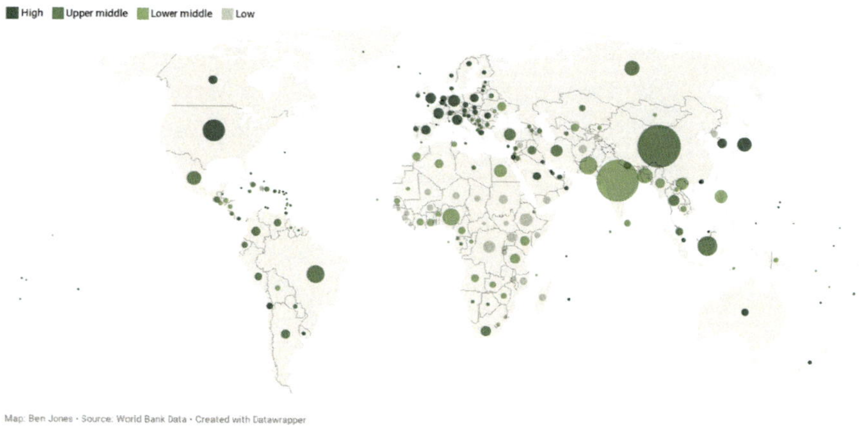

Figure 8.9. Symbol map of 2019 country population map colored by income group

These shades of green are somewhat difficult to tell apart in this map, though. The clustered arrangement leads to overlapping circles in many places, and the green saturation levels might not be quite distinct enough. In cases like this, we can use the Viridis color palette we introduced when we considered heatmaps. Figure 8.10 makes it easier to tell which countries are in which income groups, and this color encoding still expresses the notion of order, to a lesser degree perhaps.

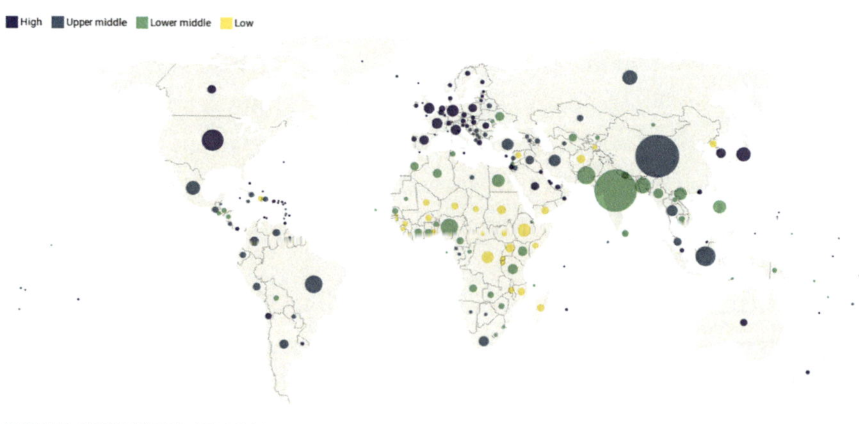

Figure 8.10. Symbol map of 2019 country population with Viridis color palette encoding income group

Of course, instead of using the color channel to encode a categorical variable like region or income group as before, we can use this fourth visual channel to encode yet another quantitative variable. For example, if a secondary interest of ours was to compare the level of urbanization of each country, we could use color saturation to encode the percent of the population living in an urban environment, as shown in Figure 8.11.

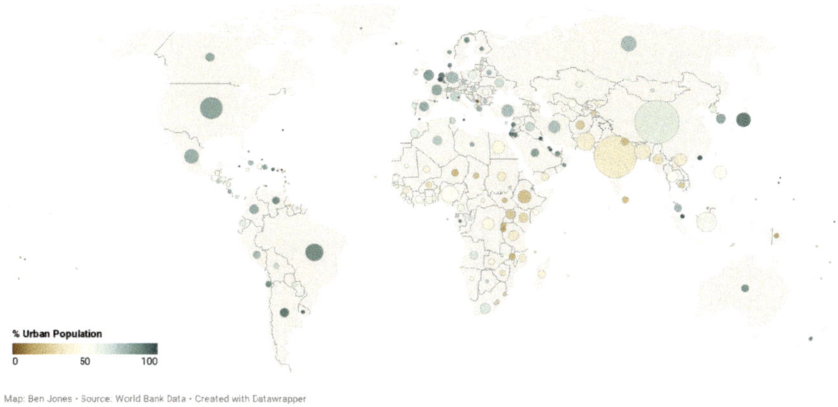

Figure 8.11. Symbol map of 2019 country population with color as % urban population

Note also that this color encoding uses a diverging color palette that changes from brown to turquoise over the range of the quantity. The center point is 50%, meaning we can quickly and easily tell which countries have more than half of their population living in a city, and which have less than half living in a city. We've been looking at this data set quite a lot by this point, and this is the first time we can actually see *where* the clusters are located on the planet.

The shapes on symbol maps don't always have to be circles. They can be squares or triangles or whatever shape we want. And we can have more than one shape on the same map. Shape can actually be a visual encoding channel. This is exactly what the map in Figure 8.12 shows.

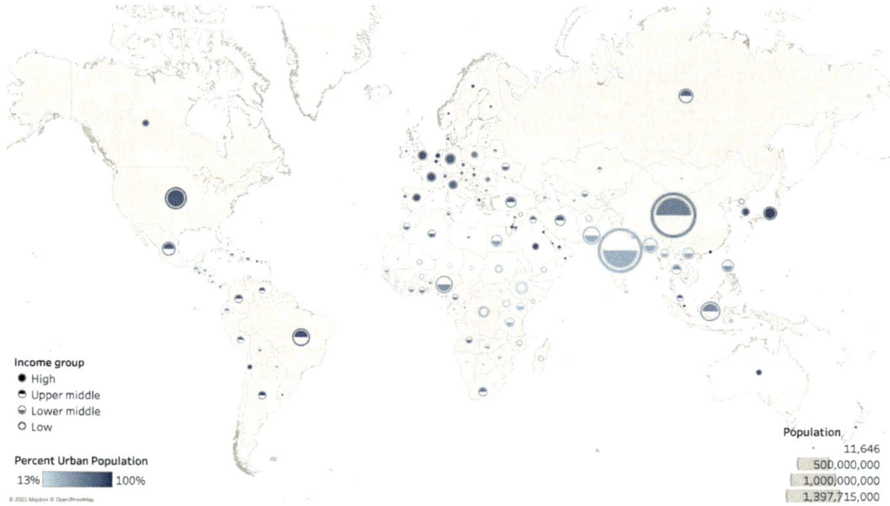

Figure 8.12. Symbol map of country population encoding shape with income group

The shape legend in the bottom left corner of the map makes the encoding clear: full circles correspond to high-income countries. Circles with only the upper half filled are upper-middle-income countries. Those with only the lower half filled are lower-middle-income countries. And finally, empty circles are low-income countries. These choices aren't arbitrary; the shapes themselves express the inherent order in this variable.

This map now shows five different variables in the same view: four quantitative variables and one categorical variable. It contains a lot of information, which can make it challenging to read. Once you spend enough time learning the encodings and orienting yourself to what's shown, you can begin to answer a great deal of questions.

This leads us to our next Chart Reading Tip:

> **TIP #16:** Charts require readers to invest varying amounts of time to learn their encodings. The value of a chart is in the ratio of what you get from it to what you must first give.

Of course symbol maps don't always have to deal with countries. They can deal with other geographic entities like states or provinces, counties, cities, and even addresses or random points of interest. But beyond that, they don't even need to be based on the Earth's surface at all.

We can imagine symbol maps that superimpose shapes on a building's floor plan, or a chess board, or a human body. Maps in general show spatial data, and maps that show position on the Earth show geospatial data. For example, we can look at the original Los Angeles professional basketball team's shot results for the 2018–2019 regular season, as shown in Figure 8.13. Yellow dots are shots that went in and purple ones are misses.

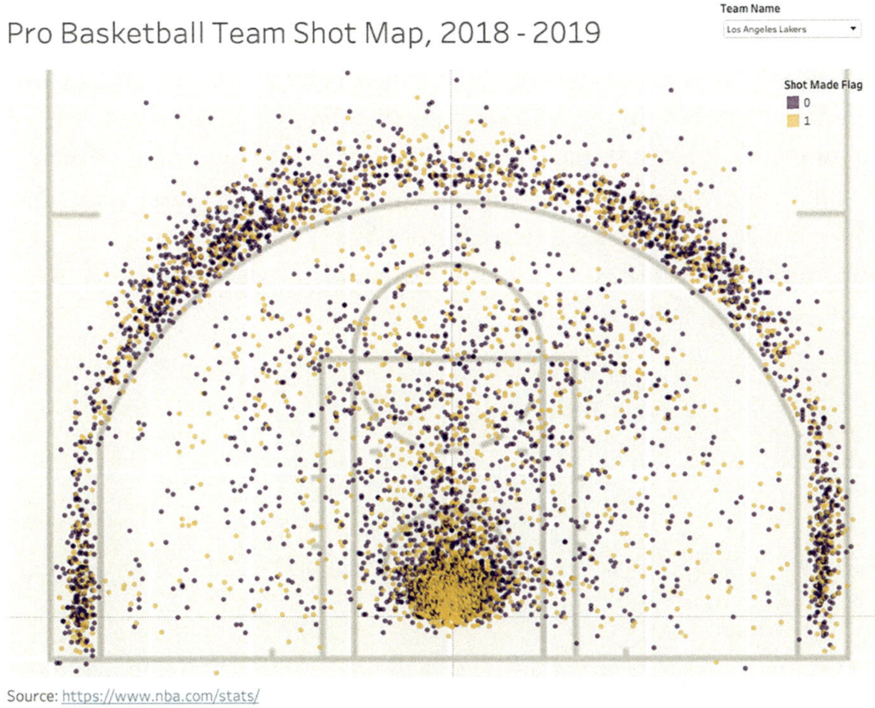

Figure 8.13. A basketball court map of shots made and missed

These maps we've considered all place a single dot, circle, or shape at a specific point on an underlying map. These symbol maps are helpful because they show us where individual data points or aggregated groups of data points were physically located when the data was collected.

The Flow Map

Often, though, we're interested in seeing more than just location; we want to see lines that express *movement* from one location to another on a map. Sure, we'd love to know where the take-out restaurant is located in town, but what we really need to know is how to get there. With the navigation devices in our pockets and on our dashboards, we probably use **flow maps**, or path maps, more often than any other type of geospatial map. Let's consider them now.

We've already actually considered flow maps, of a sort. The various map projections we considered in the first section of this chapter included graticules, lines of constant latitude and longitude that act like scaffolding for the map. Movement along these types of lines would leave one of your two coordinates unchanged, depending your direction of travel. Parallels have lines of constant latitude and meridians have lines of constant longitude.

Similarly, **topographic maps** are another form of flow map that include background layers that provide information about the territory. Topographic maps feature contour lines, which are the light brown lines shown in Figure 8.14.

Figure 8.14. A topographic map of the Enchantments of the Alpine Lakes in Washington[56]

56. Credit: U.S. Geological Survey, USGS Topo Maps, https://viewer.nationalmap.gov/basic/

These contour lines are lines of constant elevation.[57] If you were to walk along these lines, you would never have to go uphill or downhill. You'd walk along at the exact same elevation.

I had the opportunity to hike through this exact area, known as the Enchantments, in 2018. I can tell you, we did not stay at the same elevation the whole time. If I were to recount the story of this epic, four-day backpacking trip through this unforgettable part of the Alpine Lakes Wilderness in the Central Cascades of the state of Washington, I'd describe ascending Aasgard Pass, seeing the wild white mountain goats on the other side of the pass, and taking a refreshing swim in Perfection Lake.

I'd also talk to you about how far we hiked, how long it took, and how fast we had to go to make it through in time. If I had a topographic map, I'd trace our path and show you where we went, and where we stopped along the way. Luckily, I had a GPS device on me during the entire trip and so was able to create the flow map seen in Figure 8.15, which shows our path through the region over the course of the trek.

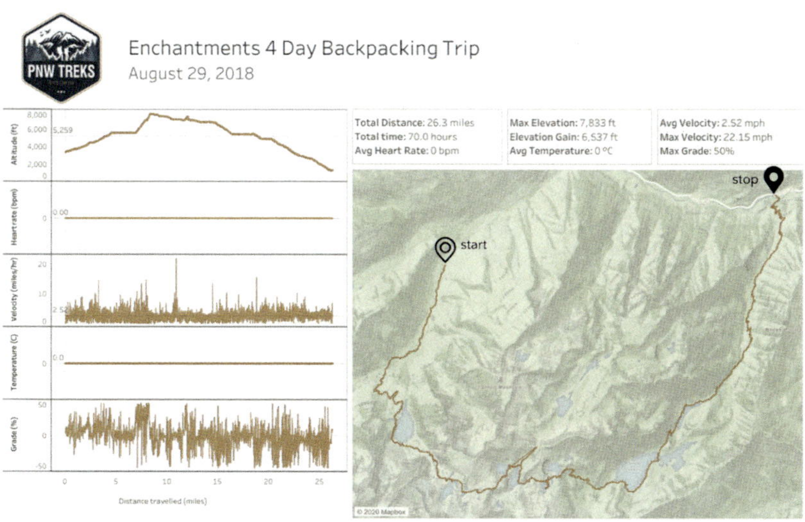

Figure 8.15. A dashboard showing the path of travel through the Enchantments

The start and end map markers convey the direction of travel, but arrows can perform this same function. Paths on maps can sometimes include

57. https://pubs.usgs.gov/gip/TopographicMapSymbols/topomapsymbols.pdf

stylized arrows that indicate starting and ending points, but not necessarily the precise path of travel. For example, a creative flow map shown in Figure 8.16 shows the flow of goods from the United States to its top fifteen trading partners based on data from the U.S. Census Bureau obtained by map maker Tony Cohen in 2011. Notice the use of the Robinson map projection, the encoding of the width of the line by trade value (in $USD billions), and the encoding of the direction of trade by use of the red-green color scheme we referred to earlier that may be difficult for some to use.

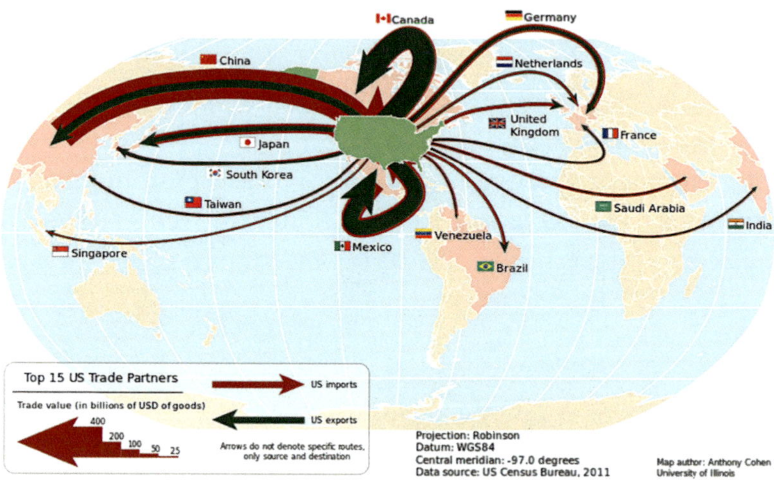

Figure 8.16. Flow map of U.S. imports and exports to its top 15 trade partners[58]

Perhaps the most famous flow map from any of the pioneers of the field of data visualization is Charles Joseph Minard's flow map shown in Figure 8.17. It depicts the disastrous campaign of Napoleon's French army that marched from Poland to Moscow in 1812. The tan band shows the approximate path of their advance toward Moscow. The black band shows the path of their retreat. Ominously, the width of the band is generally proportional to the number of Napoleon's men still alive at that point in the campaign. And the line chart at the bottom indicates the temperature along their path, shown from right-to-left as the retreat was a westward trudge through the frozen Russia winter.

58. TonyCohen, CC BY-SA 3.0 <https://creativecommons.org/licenses/by-sa/3.0>, via Wikimedia Commons

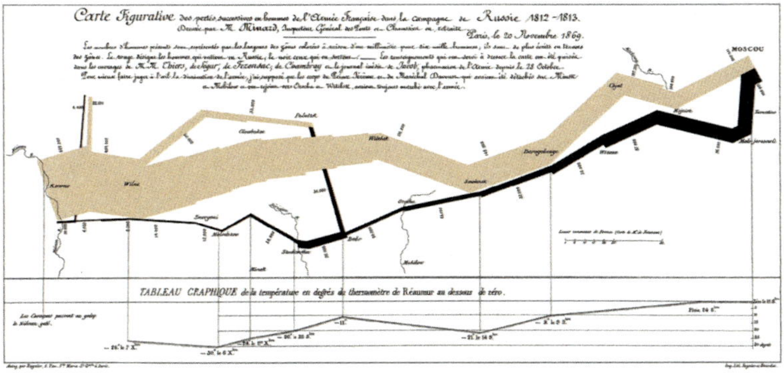

Figure 8.17. Charles Joseph Minard's 1869 flow map of Napoleon's 1812 March on Moscow[59]

Besides the direction maps we pull up on our phones almost every time we leave our homes, the most useful path map might be the subway maps (aka tube maps or transit maps) of various city subway systems around the world. These interesting maps take many liberties with the actual geography in order to convey to the subway rider the order of the stations. If you're interested in understanding the precise path of your travel, these maps won't do it. But if you only need to know which train to take, which station to exit the train, and how many stops are left, there's really nothing better.

Figure 8.16. An unofficial redesign of the MBTA subway map by Michael Kvrivishvili[60]

59. Charles Minard (1781-1870), Public domain, via Wikimedia Commons https://commons.wikimedia.org/wiki/File:Minard.png
60. Michael Kvrivishvili, CC BY 2.0, via Wikimedia Commons: https://commons.wikimedia.org/wiki/File:2013_unofficial_MBTA_subway_map_by_Michael_Kvrivishvili.png

We can only judge a map by taking into careful consideration the sorts of tasks that it was built to accommodate. This is true of charts in general. No chart can accommodate every conceivable task, and some enable us to do certain types of tasks better than others.

The Choropleth Map

The jump from symbol maps to path maps involved moving from 0D marks to 1D marks. Likewise, the jump from path maps to our next map type, the **choropleth map**, involves moving from 1D marks to 2D marks: regions or areas.

Places on the planet are often defined by closed boundaries that form distinct regions, with notions of "inside," "outside," and "on the line." These boundary lines can be natural, like the coastline that separates an ocean from the land. They can be physical, like the perimeter of a building. They can be political, like the border separating two nations, states, or counties. What they have in common is that they all define zones that are separated from other zones in some way and for some purpose.

The maps we've already considered showed us some of these boundary lines. The world maps showed both coastlines and country borders. The basketball shot map showed lines separating the different zones on the court, like sidelines and end lines separating "in bounds" from "out of bounds," and the three-point arc separating where shots are worth three points versus just two.

On a two-dimensional map projection, these closed borders form polygons. Some polygons are basic shapes, like the rectangular "key" in front of the hoop on a basketball court, and others are highly irregular shapes, like an island's coastline whose edges are jagged no matter how much we zoom in or out.

In these cases, situations arise in which we want to visualize some quantitative or statistical measure that's characteristic of the entire region because it's based on an aggregate summary of values that pertain to that region. That's a mouthful, so let's consider an example.

We may want to compare crime rates by county, which is typically found by dividing the number of reported crimes that took place within the county by the number of people who live in it. These are situations in which choropleth maps come in very handy.

The first known choropleth map was created in 1826 by French mathematician Pierre Charles François Dupin. He showed the availability of basic education by department in France by filling each department with shades of gray ranging from white to black, as shown in Figure 8.17.

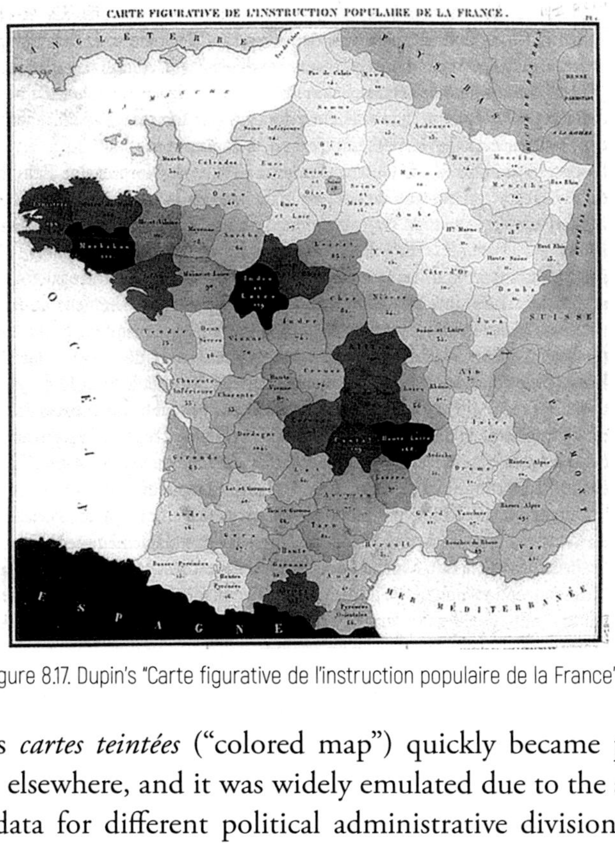

Figure 8.17. Dupin's "Carte figurative de l'instruction populaire de la France"[61]

Dupin's *cartes teintées* ("colored map") quickly became popular in France and elsewhere, and it was widely emulated due to the availability of census data for different political administrative divisions in many countries.

Today, companies use these types of maps to track revenue by sales territory, news organizations use them to cover election results by state or county, and government agencies use them to publish high school graduation rates, among many other scenarios.

61. Source: Wikimedia Commons, Public Domain: https://en.wikipedia.org/wiki/File:Carte_figurative_de_l%27instruction_populaire_de_la_France.jpg

If we return to our example of urban population by country, we can learn a valuable lesson about choropleth maps. A world map with countries shaded based on the size of the *urban population* looks very, very similar to a map with countries shaded by *total population*, as shown on the left side of Figure 8.18. Filling in choropleth shapes based on the totals of a population-dependent variable often gives little else than the information contained in a population map of the same area.

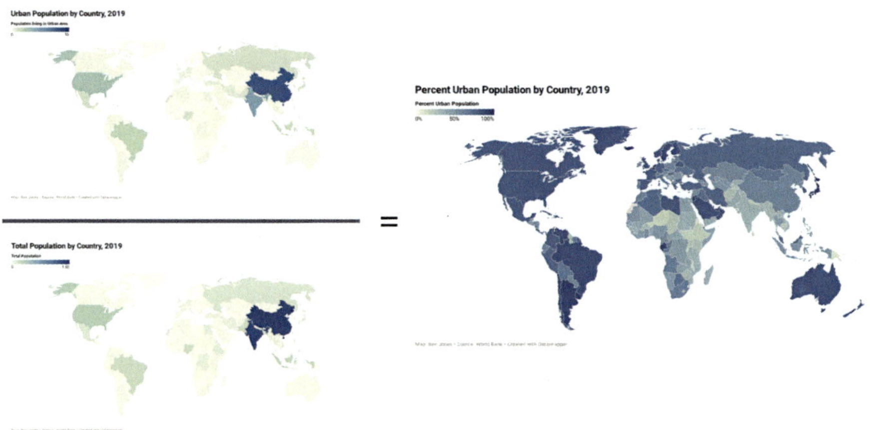

Figure 8.18. How normalizing a variable results in more than just a population map

For this reason, it's customary and advisable to **normalize** quantities used to encode color saturation on choropleths. We use them to view *rates or percents*, as opposed to *totals*. To normalize a quantitative variable, we divide each area's value by its population. In this case, dividing each country's urban population by its total population gives us the fraction of the population that lives in an urban environment. Multiplying that number by 100 gives us the percent. A choropleth map of this normalized variable is shown on the right side of Figure 8.18. This looks nothing like the population map on the bottom of the left side—the denominator, if you will—and it actually adds a new perspective about the topic.

To conclude this section on the choropleth map, let's consider a real-world case study that combines a number of different concepts we've covered in the book. This case study will allow you to put your new skills as an experienced chart reader to the test. I don't mean to disparage the person who created the map referenced here, and we should be careful to avoid snark or ridicule when we critique the work of others. We all make mistakes, and chart-makers are in no way exempt from this reality. It could be due to software defaults, or requirements imposed on us by others, or just plain human error. To err is human.

With that important mindset, let's turn our attention to U.S. public high school graduation rates. I have three teenage boys who all attended high school in the state of Washington in the year of writing this book. If we want to understand how graduation rates are different by state in the U.S., we can turn to a report by the U.S. National Center for Education Statistics titled "The Condition of Education 2020."[62] This report features the choropleth map shown in Figure 8.19. It's normalized; it shows graduation rates instead of the raw number of graduates by state. And while that's great, take a closer look at this map and see if you can find the problem with it. Try to channel your inner Mary Eleanor Spear, who taught us in the introduction that "there is quite a difference between simply looking at a chart and seeing it."

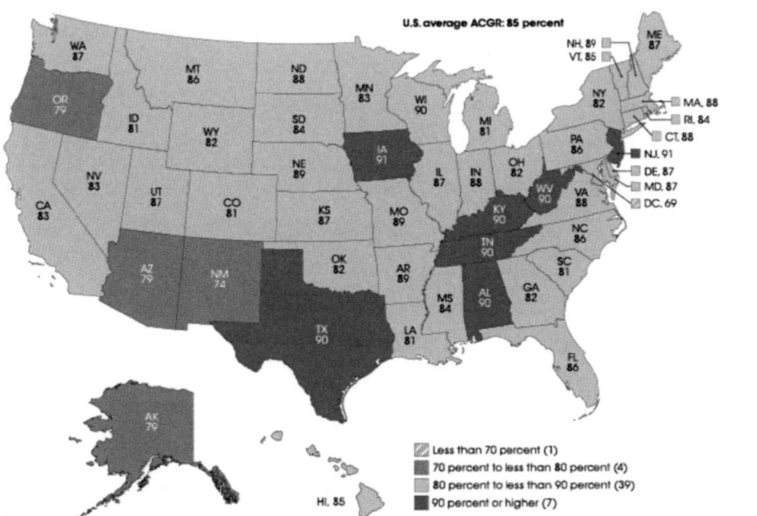

NOTE: The ACGR is the percentage of public high school freshmen who graduate with a regular diploma within 4 years of starting ninth grade. The Bureau of Indian Education and Puerto Rico are not included in the U.S. average ACGR. The graduation rates displayed above have been rounded to whole numbers. Categorizations are based on unrounded percentages.
SOURCE: U.S. Department of Education, Office of Elementary and Secondary Education, Consolidated State Performance Report, 2017–18. See *Digest of Education Statistics 2019*, table 219.46.

Figure 8.19. A choropleth map of public high school graduation rates by state, 2017–18

62. https://nces.ed.gov/pubs2020/2020144.pdf

If the error eludes you, don't worry; it did for me, too, along with other data visualization experts I know. It may help if I pose a specific question to you: which state had a higher graduation rate in 2017–18, Oregon (OR) or Washington (WA)? These states are neighbors to each other in the Pacific Northwest corner of the country. At a first glance, you may be tempted to answer that Oregon does; after all, its state shape is colored a darker shade of green than Washington's. But when you look closely at the state labels, you see that Washington (87%) actually had a higher graduation rate than Oregon (79%) that year. What gives?

If you examine the legend, you'll see that it breaks the expressiveness principle we learned back in the very first chapter of this book. Recall that this principle, according to Tamara Munzner, states that "the visual encoding should express all of, and only, the information in the dataset attributes." In this case, the encoding is color saturation: various levels of saturation of a forest green color correspond to different groups, or bins, of graduation rates. But the choice of saturation levels expresses an order that doesn't exist in the data. States with a graduation rate between 70% and 80% have a darker shade of green than those with a graduation rate between 80% and 90%. How should we revise this map?

Let's start by getting rid of the color saturation bins that we see in the legend. The only reason to create a "stepped" color ramp like this is if the bins are meaningful for some specific reason similar to the way grades are assigned at school, the very topic of the chart. For example, if states get federal funding based on predefined graduation rate groups, then we would want to use cutoff points for the color saturation levels in this way. And while that may be the case, you and I are simply interested in comparing relative rates, so we'll encode color saturation based on the full range of the quantity itself, with no bins. Doing this results in the choropleth map shown in Figure 8.20.

Notice that the state shape for Washington is now appropriately darker than the shape for Oregon. It's also a darker shade of green than the shape of its other neighbor, Idaho (ID), which wasn't the case in the original due to the rate bins. Idaho's graduation rate (81%) was much closer to Oregon's (79%) than Washington's (87%) that year, and now color saturation more accurately represents that.

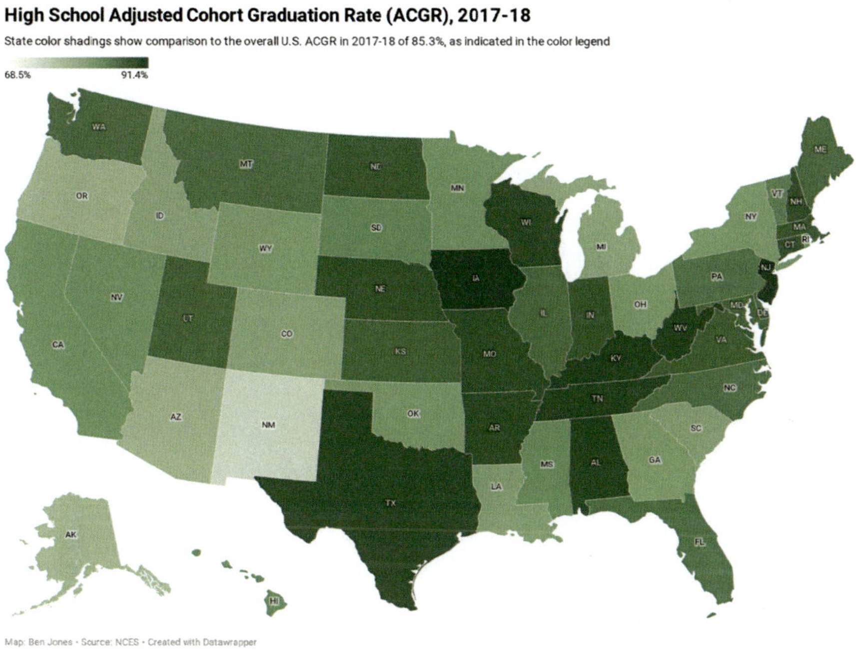

Figure 8.20. A remake of the state gradual map with a corrected color map

As we've seen, sometimes we want to compare values to an interesting point that lies in between the minimum and maximum values. What if we were interested in knowing which states were above the national graduation rate, and which were below the national graduation rate that year? It's very difficult to tell using our updated choropleth map in Figure 8.20. That kind of question is better answered with a diverging color map, as shown in Figure 8.21.

As we can see by examining the color legend in the top left corner, the color hues diverge from brown to turquoise, and they switch hues at a graduation rate of 85.3%. Why this seemingly arbitrary pivot point, though? If we read the sentence below the title of the map, we'll see that this exact value corresponds to the overall high school graduation rate in the United States in the 2017–18 school year. With this value as the diverging color scale's pivot point, it's easy to tell on which side of the national rate each state fell that year.

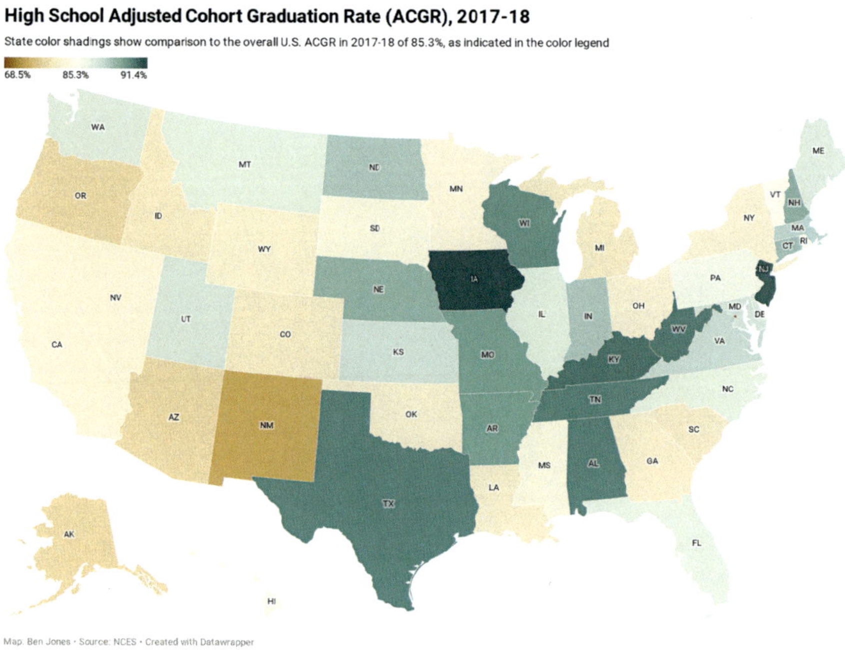

Figure 8.21. A choropleth with a diverging colors for comparison with a middle value

We can also quickly tell which state had the highest graduation rate: it was Iowa (IA) in the middle of the country that claimed the maximum value with a graduation rate of 91.4%. We can tell because it's easy to see that it has the darkest shade of turquoise on the map. Only New Jersey (NJ) comes close (90.9%).

But which had the lowest graduation rate? At first, it seems like perhaps New Mexico has the most saturated brown shape on the map, but its graduation rate was 73.9%, not quite as low as the lowest value of 68.5%. So which area had the lowest graduation rate, then? It's very difficult to see on the map, and it's actually not a state at all. It's Washington, D.C., which had a graduation rate of 68.5%. The darkest brown shape is actually just a small dot right below the MD label (Maryland), roughly in the middle of the Eastern seaboard.

Incidentally, we might be tempted to use the popular red-blue diverging color palette to show these state graduation rate figures as shown in Figure 8.22. It feels very "American," and we may feel that we win some patriotic points with this color choice. But I would advise against using any red-blue color scheme on a map of the United States that doesn't show political party data. For many readers, this color choice will express political affiliation, but that's not the topic of the chart at all. These two colors are closely associated with the largest two political parties in the country, and many will assume that they're looking at a map of election results.

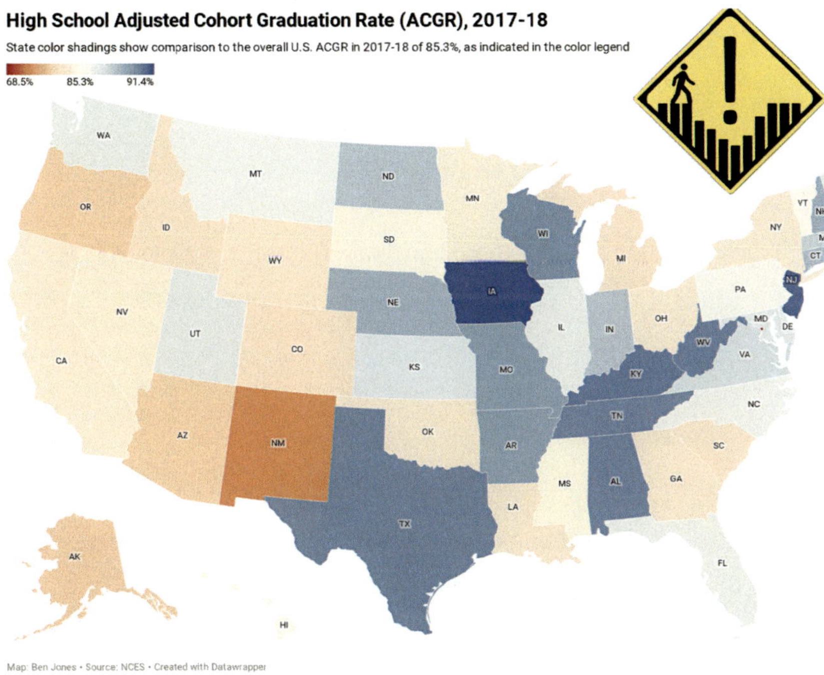

Figure 8.22. A potentially misleading use of red-blue colors to show nonpolitical data

Of course this restriction would only apply when showing data to an audience that's intimately familiar with the spectrum of U.S. politics. Similarly, many different audiences will have ingrained associations with different colors and shapes that the chart maker should take into account, and only contradict when there is a strong reason to do so.

With that, we wrap up our section on maps. The field of cartography is incredibly deep, complex, and rich; it's an entire discipline in and of itself. While this chapter did not turn you into an expert in maps, my hope and belief is that it gave you a much better understanding of how to read maps that feature data in the form of points, lines, and shapes. This is a critical ability for the would-be chart interpreter, as data often contains information tied to the places we live, work, and visit.

As the late Lord Harold Samuel said in 1944 upon founding one of the United Kingdom's largest property companies, "Location, location, location."

FURTHER LEARNING

Video: "Why all world maps are wrong" by Johnny Harris of Vox Video Lab (December 2, 2016: https://www.youtube.com/watch?v=kIID5FDi2JQ

Rhumb Lines and Map Wars: A Social History of the Mercator Projection by Mark Monmonier (University of Chicago Press, 1st ed. October 1, 2004): https://www.amazon.com/Rhumb-Lines-Map-Wars-Projection/dp/0226534316/

Interactive Mercator Map, "The True Size", created by James Talmage and Damon Maneice: https://thetruesize.com/

"Projection Transitions" Observable notebook by Mike Bostock: https://observablehq.com/@d3/projection-transitions

Video: "The genius of the London Tube Map" by Michael Bierut (November 3, 2018): https://www.youtube.com/watch?v=iBErp8qvWZg

Resource: "What to consider when creating choropleth maps", Datawrapper Academy: https://academy.datawrapper.de/article/134-what-to-consider-when-creating-choropleth-maps

Conclusion

WITH NEW EYES TO SEE

"The voyage of discovery is not in seeking new landscapes but in having new eyes."
—Marcel Proust

And here we are at the end of a voyage of discovery. We have discovered much about the visual language of data and we have looked closely at many different chart types and their various configurations. From bar charts to line charts to treemaps, these visualizations present richly textured landscapes to explore and curiously coded messages to interpret.

But we have discovered even more about the way our own eyes and mind work when they encounter these graphical forms. We learned about ways that charts can confuse and mislead us. Our "data pitfall alert" sign flagged us to these graphical gaffes: bar charts with truncated axes, pie charts that don't add up to 100%, line charts that include partial time periods, and a handful of other such "humdingers" as my mother Jane would call them.

We examined the nature of our initial reactions to looking at these charts. And we challenged ourselves to go beyond merely looking at them so that we can actually *see* them. This was the admonition of our muse, Mary Eleanor Spear. Her quote that we considered in the introduction of the book is worth repeating. This time ask yourself if you feel more confident in your ability to carry it through:

> *Learn to see details. There is quite a difference between simply looking at a chart and seeing it. Looking is your first visual impression, while seeing involves the studying of distinct parts of the visual.*

In order to help us see at a deeper level, we took time to note down specific chart reading tips. Here is the full list of 16 tips compiled into one place for ease of reference:

16 Chart Reading Tips:

1. Consider what data is shown in the display, and what relevant data might not be shown.
2. Figure out which variables in the data encode which visual channels in the chart.
3. If there are continuous axes, notice where they start and stop.
4. Ask yourself whether the chart's encodings and design help you answer your most important questions.
5. If the chart expresses completeness or fullness, make sure the segments are MECE: mutually exclusive and collectively exhaustive.
6. If a chart expresses change over time, check which direction corresponds with increasing time. It's likely left to right, but might not be.
7. If a chart expresses value or worth, check which direction corresponds with "more," "better," or "improvement." It's likely up, but might not be.
8. When data shown on a chart has been grouped by time period, verify whether the time periods are all complete, because some of them may only be partially complete.
9. Identify what brand new questions the chart prompts in your mind, and consider what further information is needed to answer them.
10. When using charts to gain understanding about a topic, try looking at the same quantities and categories using a variety of different encodings and arrangements.

11. Consider how the data in the chart has been aggregated (sum, average, median, etc.), and whether it's suitable to answer your most important questions.
12. If there are interesting values in the chart such as outliers, be sure to ask for labels or annotations to help you identify them.
13. Find out the operational definitions of the variables shown in the chart, learn how the records were collected, and consider what potential caveats these details introduce.
14. If a model like a regression line has been applied to the data in the chart, determine what type of model it is, and how well it fits the data.
15. Pay close attention to the different legends in the chart and see how they reveal encodings other than position such as size, color, or shape.
16. Charts require readers to invest varying amounts of time to learn their encodings. The value of a chart is in the ratio of what you get from it to what you must first give.

What's Next?

Now that you're more confident reading charts, graphs, and maps that others have created, or that software defaults have placed in front of you, it's time to learn how to roll up your sleeves and work with the raw data itself. Nothing replaces diving into the data, exploring its contours, spinning it around, and looking at it from another direction, finding its flaws, and crunching the numbers for yourself. As long as we let others do all of the data preparation and analysis for us, we run a much greater risk of being misled or deceived.

That's why the next book and course in this series, *From Data to Wisdom*, is designed to give you the skills you need to extract insight from data that others haven't spoon-fed to you. Ultimately, you own the process of developing the wisdom you need to make better decisions. Becoming competent at data analysis is a necessary part of that process in our era.

Made in the USA
Las Vegas, NV
03 April 2024

88185903R00133